T0211786

Control Systems

D. Sundararajan

Control Systems

An Introduction

 Springer

D. Sundararajan
Formerly at Concordia University
Montreal, QC, Canada

ISBN 978-3-030-98447-2   ISBN 978-3-030-98445-8   (eBook)
https://doi.org/10.1007/978-3-030-98445-8

This Springer imprint is published by the registered company Springer Nature Switzerland AG
The registered company address is: Gewerbestrasse 11, 6330 Cham, Switzerland

Preface

A control system is composed of a set of components to produce a desired response for a given input and widely used in several areas of science and engineering. A basic continuous-time control systems course is usually offered to undergraduate students in electrical, mechanical, mechatronic, chemical, civil, and aerospace departments. As control system theory is important in several disciplines, modeling of the various type of systems is required. Further, the derivation of the electrical analog of other types of systems is also required, as analysis and design can be carried out using the vast and well-developed electrical circuit and linear systems theory. Although the analysis and design is carried in the analog domain, due to the advances in digital system technology and fast numerical algorithms, systems can be implemented using digital components by transforming system models to the digital domain.

As the mathematical content of control systems is quite high, students have to be made comfortable in their learning of this subject through a number of appropriate examples, figures, and programs. Further, it has to be pointed out how the theoretical analysis is approximated numerically to include practical considerations in applications. In conjunction with theoretical analysis and a laboratory class, programming is essential for getting a good understanding of the subject. The course should end with one or two good projects of complexity that the students can handle. Further, each student should practice the basic concepts, such as matrix analysis and Laplace transform, with paper and pencil and programming as much as necessary for their good understanding. The essentials of control systems are linear system theory, transform methods and their computational aspects, and control systems analysis, design, and implementation.

This book is primarily intended to be a textbook for an introductory course in continuous-time control systems for senior undergraduate and first-year graduate students in several engineering departments. It can also be used for self-study and as a reference. The prerequisites for studying this subject are first courses in linear algebra, calculus, mechanics, circuit theory, signals and systems, and basic programming.

The features of this book are the detailed coverage of basic principles of control systems with MATLAB® programs (available online); clear, concise, and simplified presentation of the difficult concepts using transform theory; physical explanation of concepts; large numbers of figures and examples; and clear, concise, and, yet, comprehensive presentation of the topics. Emphasis on physical simulation of systems is a unique feature of the book, making it easier to understand system behavior.

Answers to selected exercises marked * are given at the end of the book. A solutions manual and slides are available for instructors at the website of the book. I assume the responsibility for all the errors in this book and would very much appreciate receiving readers' suggestions and pointing out any errors (email:d_sundararajan@yahoo.com). I am grateful to my editor and the rest of the team at Springer for their help and encouragement in completing this project. I thank my family for their support during this endeavor.

D. Sundararajan

Contents

Abbreviations

BIBO	Bounded-input bounded-output
DC	Direct current, sinusoid with frequency zero, constant current or voltage
Gm	Gain margin
Im	Imaginary part of a complex number or expression
LHP	Left-half of the s-plane
LTI	Linear time-invariant
RHP	Right-half of the s-plane
PD	Proportional-derivative
PI	Proportional-integral
PID	Proportional-integral-derivative
Pm	Phase margin
Re	Real part of a complex number or expression
ROC	Region of convergence
SFG	Signal-flow graph
SNR	Signal-to-noise ratio

Chapter 1
Introduction

There are certain activities, which we require in our daily lives. For example, we have to heat the water to take bath. Before the advent of control systems, we did it manually by using firewood. That requires time and effort, whereas, nowadays, we set the desired temperature in an electric water heater and turn it on. The task is done automatically without further human effort. The use of control systems is widespread in our homes as well as in industries. Control systems carry out the task automatically in the most efficient manner. In this chapter, we just introduce control systems first. One of the important tasks in control system design is to test that its performance is as required. For that purpose, some standard signals are used, while the actual signals in control systems have arbitrary amplitude profile.

1.1 Basics of Control Systems

A system carries out some task in response to an input signal. Control system is an interconnection of components, such as the controller, actuator, and plant, to produce a desired response. For example, an electric motor delivers mechanical rotational power when we energize it with electrical power. Apart from large number of industrial applications, control systems are often used for our comfort in our homes, such as room temperature control, water heater control, and voltage stabilizers for voltage control. An open-loop control system is a system with a controller and actuator to provide a desired response without any feedback, as shown in Fig. 1.1. Open-loop control system is like driving a car in a zigzag road with our eyes closed (without feedback from our eyes). A closed-loop control system is a system with a controller and actuator to provide a desired response with some feedback, as shown in Fig. 1.2. The controller produces the control signal. Its function is to hold the desired response at a desired value regardless of the changing environment around it. Closed-loop control system is like driving a car in a zigzag road with our eyes open (with feedback from our eyes).

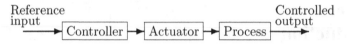

Fig. 1.1 Block diagram of an open-loop control system

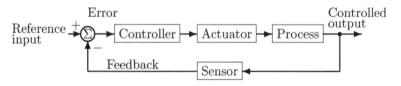

Fig. 1.2 Block diagram of a closed-loop control system

Room Temperature Control

A control system is used to maintain a desired temperature in the room, regardless of the changing temperature environment around it. In the open-loop control system, we set a desired temperature in the controller. It applies continuous heating with no regard to the actual output temperature. This type of system may not be satisfactory, at both day and night time, as the outside temperature varies considerably. In the closed-loop control system, the controller monitors the temperature in the room and if it goes high or low from the desired value, it sends a signal to activate the heater in such a way to reduce the error signal. The feedback input value to a temperature controller is an electric signal proportional to the temperature.

As control systems are used in several different applications, a transducer that transfers signals generated in one domain to another domain is often used. Transducers are used to sense signals such as flow, level, pressure, temperature, velocity, acceleration, etc. In particular, transducers those produce electrical output are most often used as electrical signals, which are very convenient for analysis, design, and simulation of systems. A tachometer is a transducer that provides a proportional voltage to the magnitude of the angular velocity of a shaft. Another commonly used transducer is the potentiometer. It is a resistor with three terminals, the third being an adjustable one. It is an electromechanical transducer that senses a mechanical displacement and provides a corresponding electrical signal. A thermistor is a transducer that is an electrical resistor, the resistance of which varies rapidly in a known manner with temperature. A thermocouple is also a temperature transducer that produces an electric current that is proportional to the temperature.

Tank Liquid (for Example, Water) Level Control

A common application of the control system is water tank level control. The control system prevents overflow of water, when the tank is full. When the water level goes down, the float also goes down and the attached valve opens the water inlet to the tank. When the water level reaches the set value, the float goes up and the attached valve closes the water inlet to the tank.

Human Body Control

The human body is an ideal model of a control system, as well as in other applications of science and engineering. In practical applications, we try to emulate the way biological systems carry out various tasks. For example, our eyes adjust the gain to suit light levels and normalize responses to contrast. In image processing applications, we design algorithms to emulate visual systems. Some of the targets for control are:

- Maintenance of blood pressure and body temperature within the sustainable range
- Maintenance of water, salt, and electrolytes
- Discharge of waste materials from the body
- Closing of the eyelids, when an object approaches the cornea of the eye

Automatic Gain Control

Automatic gain control is a closed-loop control system in an amplifier to maintain a suitable signal amplitude at the output, despite wide variations in the signal amplitude at the input. A typical application is radio receivers. The gain control is necessary to adjust for the different amplitudes of the signals received from different radio stations. Further, gain control is required to receive the signal even from a single station due to fading because of atmospheric disturbances and other reasons. The gain control system reduces the volume for strong signals and raises it when the signal is weak.

Frequency Control in Electrical Power System

The frequency and voltage of the electrical power supply that is distributed to industries and homes have some specifications to comply with. That is, the frequency and voltage variations must be within some prescribed limits. A drop in speed of the generator due to increased load causes the control system to increase the input (admit more steam into the turbine in the case of steam power plants) resulting in an increase in the speed of the generator. This frequency control brings out the basics of closed-loop control systems.

1. Sensing of the actual response.
2. The sensor output is interpreted in terms of a deviation from a control point.
3. Necessary corrections are applied to restore the response to its desired form.

Basic Terminology Used in Control Systems

Actuator   Device that provides motive power to the process

Comparator   Computes the difference between the desired and actual output

Controller   Device that computes the control signal. Composed of comparator and compensator

Output   The output of the system to be controlled

Plant   Combination of the actuator and the system under control

Process   System or device, whose output is to be controlled

Reference input   The desired output of the system. Also called set point or input

Sensor   Device that detects and measures some physical effect and generates a
  signal proportional to the effect
Transfer function   Ratio of the transforms of system output and input. That is,
  the transform of the input multiplied by the transfer function is the output in the
  transform domain

1.2   Basic Signals

Signals are a source of information, such as an audio or a temperature signal. As
signals vary with time or some other independent variable, they are also referred
as functions. For example, $x(t) = \sin(t)$, where $x(t)$ is its amplitude and t is
the independent variable. The amplitude profiles of practical signals and those
of the responses of practical systems are arbitrary. Therefore, signals have to be
decomposed in terms of some well-defined basic signals, such as the impulse
and sinusoid, for compact representation and easier processing. Systems can be
characterized by their responses to the basic signals, impulse, unit-step, ramp,
parabola type, and sinusoids. Then, using these responses, the system response
for arbitrary signals can be easily determined by decomposing the input signal in
terms of basic signals and summing the responses to all the components, assuming
linearity property of linear systems. The basic signals are used as intermediaries in
the analysis of signals and systems. They are not practical signals. For example, the
duration of the sinusoid is infinite. The bandwidth of impulse is infinite. However,
they can be approximated, in practice, to a required accuracy. The response of
systems to them can be measured, in practice, with acceptable tolerance. While the
responses of test signals of a system are related, the unit-step signal is the easiest to
generate and is usually used for performance tests.

1.2.1   The Unit-Step Signal

On the interval it is defined, a function is continuous if a pen can trace its graph
completely without being lifted from the page. While a sinusoid is a continuous
function, the unit-step signal is discontinuous. The step signal is indispensable in
the analysis of systems, for example, in modeling a switch. For this purpose, a
generalized function, called the impulse, is defined to make the unit-step signal its
integral. In control systems, the specifications of a system are usually given in terms
of its unit-step response. The unit-step function $u(t)$, shown in Fig. 1.3a, is defined
as

$$u(t) = \begin{cases} 1 \text{ for } t \geq 0 \\ 0 \text{ for } t < 0 \end{cases}$$

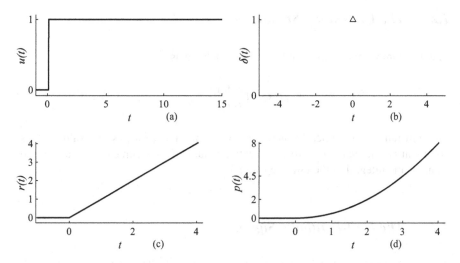

Fig. 1.3 (**a**) The unit-step signal, $u(t)$; (**b**) the unit-impulse signal, $\delta(t)$; (**c**) the unit-ramp signal, $r(t)$; (**d**) the unit-parabolic signal, $p(t)$

The unit-step signal has a value of one for positive values of its argument t and its value is zero otherwise. The delayed version of the unit-step signal is defined as

$$u(t - a) = \begin{cases} 1 \text{ for } t \geq a \\ 0 \text{ for } t < a \end{cases}$$

1.2.2   The Unit-Impulse Signal

Although step and impulse functions are of fundamental importance in signal and system analysis, there is no analytic way to define these functions, as they are not analytic. A simple, rigorous, and usable interpretation for the derivative of a function at a point of discontinuity is given by defining a generalized function—the impulse. A regular function is usually defined by its value. The impulse function is called a generalized function, since it is defined by the result of its operation (integration) on an ordinary function. The continuous unit-impulse signal $\delta(t)$, located at $t = 0$, is defined, in terms of an integral, as

$$\int_{-\infty}^{\infty} x(t)\delta(t)\, dt = x(0)$$

assuming that $x(t)$ is continuous at $t = 0$ (so that the value $x(0)$ is unique). The value of the function $x(t)$ at $t = 0$ has been sifted out or sampled by the defining operation. The signal is shown in Fig. 1.3b.

1.2.3 The Unit-Ramp Signal

The unit-ramp signal, shown in Fig. 1.3c, is defined as

$$r(t) = \begin{cases} t & \text{for } t \geq 0 \\ 0 & \text{for } t < 0 \end{cases}$$

The unit-ramp signal linearly increases, with unit slope, for positive values of its argument and its value is zero for negative values of its argument. The unit-ramp signal is the integral of the unit-step signal.

1.2.4 The Unit-Parabolic Signal

The graph of any quadratic function is called a parabola. The unit-parabolic signal, shown in Fig. 1.3d, is defined as

$$p(t) = \begin{cases} \frac{t^2}{2} & \text{for } t \geq 0 \\ 0 & \text{for } t < 0 \end{cases}$$

The denominator 2 makes its Laplace transform $1/s^3$. The unit-parabolic signal quadratically increases for positive values of its argument and its value is zero for negative values of its argument. The unit-parabolic signal is the integral of the unit-ramp signal.

1.3 Sinusoids

A general sinusoidal waveform is a linear combination of trigonometric sine and cosine waveforms or shifted sine and cosine functions. Sinusoidal representation of signals is indispensable in the analysis of signals and systems for the following most cogent reasons. The steady-state waveform, due to an input sinusoid, in any part of a linear system, however, complex it may be, is also a sinusoid of the same frequency as that of the input differing only in its amplitude and phase. No other periodic waveform has this distinction. Linear systems can be modeled by a linear differential equation with constant coefficients and the particular integral of the equation is a sinusoid when the input is sinusoidal. The sum of any number of sinusoids having the same frequency but arbitrary amplitudes and phases is also a sinusoid of the same frequency. No other periodic function can lay claim to this property either. The integral and derivative of a sinusoid is again a sinusoid of the same frequency. Due to these properties, system models, such as differential equation and convolution, reduce to algebraic equations for a sinusoidal input for linear systems. Further, due to the orthogonal property, any practical signal, with arbitrary amplitude profile, can be decomposed into a set of sinusoids using fast algorithms. Physical systems also, such as a combination of an inductor and a capacitor, produce an output of sinusoidal nature. The motion of a simple pendulum is approximately sinusoidal.

1.3.1 The Polar Form of Sinusoids

There are two forms of representation of real sinusoids. At a given angular frequency ω, a sinusoid is characterized by its amplitude A and its phase θ (called the polar form) or by the amplitudes of its sine and cosine components (called the rectangular form). Signal amplitude can be either positive or negative.

The polar form of a sinusoid is

$$x(t) = A\cos(\omega t + \theta), \qquad -\infty < t < \infty$$

A sinusoidal waveform has a positive peak and a negative peak in each cycle. The distance of either peak of the waveform from the horizontal axis is its amplitude A. The cosine function is periodic,

$$\cos(\omega t) = \cos(\omega t + 2\pi)$$

Any function defined on a circle will be a periodic function of an angular variable ω. It repeats its values for $t = t + T = t + 2\pi/\omega$, where T is its period in seconds. Then, the cyclic frequency of the sinusoid is $f = 1/T$ Hz (cycles/second). The independent variable t, while time in most applications, can be anything else also, such as distance.

Sinusoids $x(t) = 2\cos(\frac{2\pi}{8}t)$ and $x(t) = \sin(\frac{2\pi}{8}t)$ are shown in Fig. 1.4a. Cosine and sine waveforms are special cases of a sinusoidal waveform. Cosine waveform has its peak value 2 at $t = 0$. Taking it as a reference, its phase is defined as

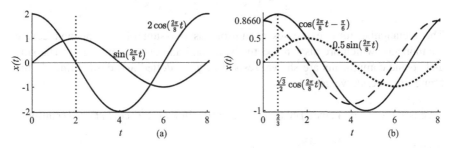

Fig. 1.4 (a) $x(t) = \sin(\frac{2\pi}{8}t)$ and $x(t) = 2\cos(\frac{2\pi}{8}t)$; (b) $x_o(t) = 0.5\sin(\frac{2\pi}{8}t)$, $x_e(t) = \frac{\sqrt{3}}{2}\cos(\frac{2\pi}{8}t)$ and $x(t) = \cos(\frac{2\pi}{8}t - \frac{\pi}{6})$

zero radians. The radian frequency is $\omega = 2\pi/8$ radians/second. Its period is $T = 2\pi/\omega = 8$ seconds. That is, it makes one complete cycle in 8 s, as shown in the figure, and repeats indefinitely for $-\infty < t < \infty$. Its cyclic frequency is $f = 1/8$ Hz. The sine waveform $x(t) = \sin(\frac{2\pi}{8}t)$ has its peak value 1 at $t = 2$. Taking the cosine waveform as the reference, its first peak occurs after a delay of 2 s, which is one-fourth of a cycle in the period 8. Since one complete cycle corresponds to 2π radians or $360°$, its phase is defined as $-\pi/2$ radians or $-90°$. That is,

$$\sin\left(\frac{2\pi}{8}t\right) = \cos\left(\frac{2\pi}{8}t - \frac{\pi}{2}\right)$$

Therefore, given a sinusoidal waveform in terms of sine waveform, it can be expressed, in terms of cosine waveform as $A\sin(\omega t + \theta) = A\cos(\omega t + (\theta - \frac{\pi}{2}))$. Similarly, $A\cos(\omega t + \theta) = A\sin(\omega t + (\theta + \frac{\pi}{2}))$. Sinusoids remain the same by a shift of an integral number of their periods, as they are periodic.

1.3.2   The Rectangular Form of Sinusoids

Figure 1.4b shows the sinusoid

$$x(t) = \cos(\frac{2\pi}{8}t - \frac{\pi}{6})$$

in solid line. Its peak value of 1 occurs at $t = 2/3$ s. Therefore, its phase is $-((2/3)/8)2\pi = -\pi/6$ radians or $-30°$. Using the trigonometric subtraction formula, we get the rectangular form as

$$\cos\left(\frac{2\pi}{8}t - \frac{\pi}{6}\right) = \cos\left(\frac{\pi}{6}\right)\cos\left(\frac{2\pi}{8}t\right) + \sin\left(\frac{\pi}{6}\right)\sin\left(\frac{2\pi}{8}t\right)$$

$$= \frac{1}{2}\sin\left(\frac{2\pi}{8}t\right) + \frac{\sqrt{3}}{2}\cos\left(\frac{2\pi}{8}t\right)$$

The rectangular form expresses a sinusoid as the sum of its sine and cosine components, which are also, respectively, its odd and even components. The sine and cosine components are shown, respectively, by dotted and dashed lines in the figure. In general, we get

$$A\cos(\omega t + \theta) = A\cos(\theta)\cos(\omega t) - A\sin(\theta)\sin(\omega t) = C\cos(\omega t) + D\sin(\omega t)$$

where

$$C = A\cos\theta \quad \text{and} \quad D = -A\sin\theta$$

The inverse relation is

$$A = \sqrt{C^2 + D^2} \quad \text{and} \quad \theta = \cos^{-1}\left(\frac{C}{A}\right) = \sin^{-1}\left(\frac{-D}{A}\right)$$

Sum of Sinusoids with the Same Frequency

An important property of the sinusoids is that the sum of sinusoids of the same frequency, but with arbitrary amplitudes and phases, is also a sinusoid of the same frequency. In order to find the sum, we have to express the sinusoids in their rectangular form and sum the respective amplitudes of the sine and cosine components. Consider the two sinusoids

$$x(t) = A\cos(\omega t + \theta) \quad \text{and} \quad y(t) = B\cos(\omega t + \phi)$$

Then,

$$z(t) = x(t) + y(t) = C\cos(\omega t + \psi) = A\cos(\omega t + \theta) + B\cos(\omega t + \phi)$$

$$= \cos(\omega t)(A\cos(\theta) + B\cos(\phi)) - \sin(\omega t)(A\sin(\theta) + B\sin(\phi))$$

$$= \cos(\omega t)(C\cos(\psi)) - \sin(\omega t)(C\sin(\psi))$$

Solving for C and ψ, we get

$$C = \sqrt{A^2 + B^2 + 2AB\cos(\theta - \phi)}$$

$$\psi = \tan^{-1}\frac{A\sin(\theta) + B\sin(\phi)}{A\cos(\theta) + B\cos(\phi)}$$

With $\theta = 0$ and $\phi = -\pi/2$ (one sinusoid being the cosine and the other being sine), the formula reduces to relation between the polar and the rectangular form of a sinusoid.

Example 1.1 Determine the sinusoid that is the sum of two sinusoids

$$x(t) = 3\cos\left(\omega t + \frac{\pi}{3}\right) \quad \text{and} \quad y(t) = 2\sin\left(\omega t - \frac{\pi}{6}\right)$$

Solution The second sinusoid can also be expressed as

$$y(t) = 2\cos\left(\omega t - \frac{\pi}{6} - \frac{\pi}{2}\right) = 2\cos\left(\omega t - \frac{2\pi}{3}\right)$$

Now,

$$A = 3, \ B = 2, \ \theta = \frac{\pi}{3}, \ \phi = -\frac{2\pi}{3}$$

Substituting the numerical values in the equations, we get

$$C = \sqrt{3^2 + 2^2 + 2(3)(2)\cos\left(\frac{\pi}{3} + \frac{2\pi}{3}\right)} = 1$$

$$\psi* = \cos^{-1}\frac{3\cos\left(\frac{\pi}{3}\right) + 2\cos\left(-\frac{2\pi}{3}\right)}{1} = \sin^{-1}\frac{3\sin\left(\frac{\pi}{3}\right) + 2\sin\left(-\frac{2\pi}{3}\right)}{1}$$

$$= 1.0472 \;\; \text{radians} = 60°$$

$$z(t) = x(t) + y(t) = \cos(\omega t + 1.0472)$$

∎

1.3.3   The Complex Sinusoids

While the sinusoidal waveform is generated by practical systems, its mathematically equivalent form, called the complex sinusoid,

$$v(t) = V e^{j(\omega t + \theta)} = V e^{j\theta} e^{j\omega t}, \qquad -\infty < t < \infty$$

is found to be indispensable for analysis due to its compact form and ease of manipulation of the exponential function. $e^{j\omega t}$ is the complex sinusoid with unit magnitude and zero phase. The complex (amplitude) coefficient is $V e^{j\theta}$. The amplitude and phase of the sinusoid is represented by the single complex number $V e^{j\theta}$, in contrast to using two real values in the real sinusoid. Due to Euler's identity, we get

$$v(t) = \frac{V}{2}\left(e^{j(\omega t + \theta)} + e^{-j(\omega t + \theta)}\right) = V \cos(\omega t + \theta)$$

The complex exponential functions separately have no physical significance. Their sum represents a physical variable, such as voltage. However, the response of a system to $V e^{j\omega t}$ yields enough information with ease to deduce the response to real sinusoids.

Real Causal Exponential Signal

Another commonly encountered signal in signal and systems is the real causal exponential signal, for example, $e^{-2t}u(t)$, shown in Fig. 1.5a.

The **time constant**, which is the inverse of the coefficient associated with the independent variable t, is 1/2. The peak value is 1 at $t = 0$. At $t = 1/2$ (one

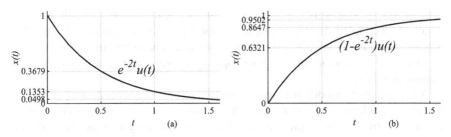

Fig. 1.5 Real causal exponential signals. (a) $e^{-2t}u(t)$; (b) $(1 - e^{-2t})u(t)$

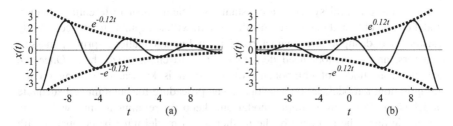

Fig. 1.6 (a) Exponentially decreasing amplitude cosine wave, $x(t) = e^{-0.12t}\cos(\frac{2\pi}{8}t)$; (b) exponentially increasing amplitude cosine wave, $x(t) = e^{0.12t}\cos(\frac{2\pi}{8}t)$

time constant), its value is $1/e \approx 0.37$. At $t = 1$ (two time constants), its value is $(1/e)^2 \approx 0.135$ and so on. Figure 1.5b shows the graph of $(1 - e^{-2t})u(t)$.

Exponentially Varying Amplitude Sinusoids

An exponentially varying amplitude sinusoid, $Ae^{at}\cos(\omega t + \theta)$, is obtained by multiplying a sinusoid, $A\cos(\omega t + \theta)$, by a real exponential, e^{at}. The more familiar constant amplitude sinusoid results when $a = 0$. If ω is equal to zero, then we get a real exponential. Sinusoids, $x(t) = e^{-0.12t}\cos(\frac{2\pi}{8}t)$ and $x(t) = e^{0.12t}\cos(\frac{2\pi}{8}t)$, with exponentially varying amplitudes are shown, respectively, in Fig. 1.6a and b.

The complex exponential representation of an exponentially varying amplitude sinusoid is given as

$$x(t) = \frac{A}{2}e^{at}\left(e^{j(\omega t + \theta)} + e^{-j(\omega t + \theta)}\right) = Ae^{at}\cos(\omega t + \theta)$$

1.4 System Modeling

Practical systems have to be modeled for their analysis and design. Further, these models have to be simulated for verification of the response. Various physical systems, such as mechanical, electrical, and hydraulic, are composed of different components. A system is called dynamic, if its present output depends on past inputs. An example is the input–output relationship of an inductor or a spring. The output varies until equilibrium condition is reached. The output of a static system depends only on current input. An example is the input–output relationship of a resistor or a dashpot.

A mathematical model of a system, that is sufficiently accurate and of lowest possible complexity, has to be first developed. Neither it is necessary nor it is possible to develop a precise model of a practical system as the required performance of engineering systems is always specified with some tolerance levels (not exact).

For many practical systems, a constant-coefficient linear differential equation model can be obtained that represents the system with acceptable tolerances. While resistors and dashpots are characterized by a linear equation, springs, masses, inductors, and capacitors need derivative terms to characterize them. Once the mathematical model of components of a system is known, depending on the configuration, a model can be developed. The procedure for obtaining an adequate model is to start with a simple model and keep on refining it until we get an acceptable one. The response of the mathematical model must be compared with simulation results for the validation of the model.

A linear time-invariant differential equation, with which we are mostly involved in this book, has an input, an output, and their derivatives in a linear combination. For example,

$$\frac{dy}{dt} + 4y = \cos(t)$$

is a first-order (contains only the first derivative of y) constant-coefficient differential equation. While we can use the differential equation model for complete analysis and design of systems, as it is difficult for practical systems, it is not used in this form. The total response of a linear system consists of a component called, zero-input response, that is due to the initial conditions alone. Another component, called zero-state response, is due to the input alone. From these responses, we can derive the transient and steady-state components. Both components are important in the analysis and design of control systems. For stable systems, the transient response decays to negligible levels in a short time and, after that, what is remaining is essentially steady-state response.

In general, the highest derivative of y present determines the order of the equation. While the first model we derive is a differential equation, it is the most difficult model to solve. Therefore, out of necessity, other models have been developed. Out of these models, we use transform and state-space models extensively in the analysis and design of control systems. The transform model uses the complex exponentials to represent arbitrary signals and system responses. The input and output of systems are related by the transfer function. While the transfer function model can be used for somewhat higher-order models and excellent for understanding the theory of linear systems, it cannot be used for practical systems, since the order involved is very high. Further, the model based on transform is applicable only for linear systems. The ultimate solution for practical use is the state-space model. This model is more general than other models with several advantages. In this model, a N-th order differential equation is decomposed into a set of N simultaneous first-order differential equations. Control system analysis and design is highly mathematical. But, one can become sufficiently proficient: (i) through practicing with paper-and-pencil method for lower-order systems and (ii) using widely prevalent software packages, such as MATLAB®, for computation and simulation of higher-order systems. In the rest of the book, we have provided both the analytical and simulation methods with numerical examples. While the state-space model will be presented in later chapters, the Laplace transform is introduced in the next chapter and, subsequently, used for system analysis and design.

1.5 Summary

- The use of control systems is widespread in our homes as well as in industries. Control systems carry out the task automatically in the most efficient manner.
- A system carries out some task in response to an input signal. Control system is an interconnection of components, such as the controller, actuator, and plant, to produce a desired response.
- An open-loop control system is a system with a controller and actuator to provide a desired response without any feedback.
- A closed-loop control system is a system with a controller and actuator to provide a desired response with some feedback.
- Transducers are used to sense signals such as flow, level, pressure, temperature, velocity, acceleration, etc.
- Signals have to be decomposed in terms of some well-defined basic signals, such as the impulse and sinusoid, for compact representation and easier processing. Systems can be characterized by their responses to the basic signals, impulse, unit-step, ramp, parabola type, and sinusoids.
- In control systems, the specifications of a system are usually given in terms of its unit-step response. The unit-step signal has a value of one for positive values of its argument t and its value is zero otherwise.
- The impulse function is called a generalized function, since it is defined by the result of its operation (integration) on an ordinary function.
- The unit-ramp signal linearly increases, with unit slope, for positive values of its argument and its value is zero for negative values of its argument.
- The unit-parabolic signal quadratically increases for positive values of its argument and its value is zero for negative values of its argument. The unit-parabolic signal is the integral of the unit-ramp signal.
- A general sinusoidal waveform is a linear combination of trigonometric sine and cosine waveforms or shifted sine and cosine functions.

- While the sinusoidal waveform is generated by practical systems, its mathematically equivalent form, called the complex sinusoid,

$$v(t) = Ve^{j(\omega t + \theta)} = Ve^{j\theta}e^{j\omega t}, \qquad -\infty < t < \infty$$

is found to be indispensable for analysis due to its compact form and ease of manipulation of the exponential function.
- Another commonly encountered signal in signal and systems is the real causal exponential signal.
- Practical systems have to be modeled for their analysis and design. Further, these models have to be simulated for verification of the response.
- The procedure for obtaining an adequate model is to start with a simple model and keep on refining it until we get an acceptable one.

Exercises

1.1 Draw the graphs of the signals $2u(t-1)$ and $3(\frac{(t-1)^2}{2})u(t-1)$.

1.2 Express the sinusoid in rectangular form. Get back the polar form from the rectangular form and verify that the given sinusoid is obtained. Find the sample values of the sinusoid for $t = 0$, $t = 2$, $t = 4$, and $t = 6$. Find the value of t nearest to $t = 0$, where the first positive peak of the sinusoid occurs?

*1.2.1 $x(t) = 2\sin(\frac{2\pi}{8}t - \frac{\pi}{6})$

1.2.2 $x(t) = 3\cos(\frac{2\pi}{8}t + \frac{\pi}{3})$

1.2.3 $x(t) = -3\cos(\frac{2\pi}{8}t - \frac{\pi}{2})$

1.3 Determine the sinusoid $c(t)$ that is the sum of the pair of given sinusoids, $a(t)$ and $b(t)$, $c(t) = a(t) + b(t)$. Find the sample values of the sinusoid $c(t)$ over one cycle starting from $t = 0$ at intervals of 1 s and verify that the sample values are the same as the sum of the sample values of $a(t)$ and $b(t)$, $a(t) + b(t)$.

1.3.1 $a(t) = \cos(\frac{\pi}{4}t + \frac{\pi}{3})$, $b(t) = -\sin(\frac{\pi}{4}t + \frac{\pi}{3})$

1.3.2 $a(t) = 2\cos(\frac{\pi}{4}t)$, $b(t) = 2\sin(\frac{\pi}{4}t)$

*1.3.3 $a(t) = 3\cos(\frac{\pi}{4}t + \frac{\pi}{3})$, $b(t) = \cos(\frac{\pi}{4}t - \frac{\pi}{4})$

1.4 Express the signal in terms of complex exponentials. Find the sample values of the two forms over one cycle starting from $t = 0$ at intervals of 1 s and verify that they are the same.

1.4.1 $x(t) = \sin(\frac{2\pi}{8}t + \frac{\pi}{6})$

*1.4.2 $x(t) = 2\cos(\frac{2\pi}{8}t - \frac{\pi}{3})$

1.4.3 $x(t) = 3\cos(\frac{2\pi}{8}t + \frac{\pi}{3})$

Chapter 2
The Laplace Transform

In the analysis of signals and systems, as in any task, we look forward, naturally, for the easiest way to solve the problem. As solving differential equations is difficult, we look forward to easier way to find its solution. For that purpose, we have to express the equation in another equivalent form, called the transformed form. For example, we can walk a few kilometers practically. For longer distances, depending on the distance, we need a car, a train, or an airplane. The transformed form is obtained by taking the transform of signals. We are already very familiar of carrying out the multiplication operation by taking the logarithm of the operands, add and find the antilogarithm to get the result of multiplication. Now, we use the same procedure in the widely used transform method of analysis of signals and linear systems. We transform the time-domain (with the independent variable time) signals to frequency-domain signals (with the independent variable frequency). The transformed signal is in every respect just as complete and specific a representation of the signal in time-domain form. The exponential signal is dominant in most of the mathematical analysis. In using logarithms to carry out multiplication, we use the real exponential signal. In the transform representation of signals, we use the complex exponential signal as basis functions to reduce the differentiation operation into much simpler multiplication operation. Then, the analysis of signals and systems reduces to algebraic operations. The complex exponential is an exponential function with a complex exponent. Basically, the signal is decomposed into complex exponentials with infinite number of complex frequencies. The operation involved in obtaining the transform representation for continuous signals is finding integrals of products of the signal to be transformed with complex exponentials. The inverse process involves complex integrals. However, the inverse can be obtained using a list of transforms of a few basic signals for most practical purposes. Sufficient practice is required to become proficient in using the indispensable transform methods.

D. Sundararajan, *Control Systems*, https://doi.org/10.1007/978-3-030-98445-8_2

2.1   Laplace Transform

The one-sided or unilateral Laplace transform $X(s)$ of the time-domain signal $x(t) = 0$, $t < 0$ is defined as

$$X(s) = \int_{0^-}^{\infty} x(t)e^{-st}dt$$

where $s = (\sigma + j\omega)$. If $\sigma = 0$, $X(j\omega)$ is the frequency response of the signal. The lower limit is 0^-, which implies that the condition of the signal immediately before $t = 0$ is taken into account. That is, any jump discontinuities or impulses at $t = 0$ are included in the analysis. Further, with this definition, handling of the initial conditions at $t = 0^-$ becomes easier. The transform exists only for values of s the defining integral converges. For all signals encountered in practical systems, the transform exists.

Example 2.1 Determine the Laplace transform of the unit-impulse signal, $\delta(t)$ from the definition.

Solution

$$X(s) = \int_{0^-}^{\infty} \delta(t)e^{-st}dt = 1, \text{ for all } s \qquad \text{and} \qquad \delta(t) \leftrightarrow 1, \text{ for all } s$$

Remember that the impulse is characterized by its unit area at $t = 0$.

Example 2.2 Determine the Laplace transform of the real exponential signal, $e^{-at}u(t)$ from the definition. Substitute $a = 0$ in the transform obtained and get the Laplace transform of the unit-step signal, $u(t)$.

Solution

$$X(s) = \int_{0^-}^{\infty} e^{-at}u(t)e^{-st}dt = \int_{0^-}^{\infty} e^{-at}e^{-st}dt$$

$$= \int_{0^-}^{\infty} e^{-(s+a)t}dt = -\frac{e^{-(s+a)t}}{s+a}\Big|_{0^-}^{\infty} = \frac{1}{s+a} - \frac{e^{-(s+a)t}}{s+a}\Big|_{t=\infty}$$

For the last limit to converge to zero, the real part of $(s + a)$ must be greater than zero. Hence, the convergence condition is $\text{Re}(s) > -a$. The Laplace transform pair for the exponential signal becomes

$$e^{-at}u(t) \leftrightarrow \frac{1}{s+a}, \quad \text{Re}(s) > -a$$

For complex-valued a, the convergence condition is $\text{Re}(s) > \text{Re}(-a)$.

Substituting $a = 0$, we get the transform pair for the unit-step signal $u(t)$ as

$$u(t) \leftrightarrow \frac{1}{s}, \quad \text{Re}(s) > 0$$

∎

With $a = \mp j\omega_0$,

$$e^{j\omega_0 t} u(t) \leftrightarrow \frac{1}{s - j\omega_0}, \quad \text{and} \quad e^{-j\omega_0 t} u(t) \leftrightarrow \frac{1}{s + j\omega_0}, \quad \text{Re}(s) > 0$$

Then,

$$\sin(\omega_0 t) u(t) = -0.5 j (e^{j\omega_0 t} u(t) - e^{-j\omega_0 t} u(t)) \leftrightarrow \frac{\omega_0}{s^2 + \omega_0^2}, \quad \text{Re}(s) > 0$$

$$\cos(\omega_0 t) u(t) = 0.5 (e^{j\omega_0 t} u(t) + e^{-j\omega_0 t} u(t)) \leftrightarrow \frac{s}{s^2 + \omega_0^2}, \quad \text{Re}(s) > 0$$

2.1.1 Properties of the Laplace Transform

Instead of using the transform definition for solving all the problems, the use of properties makes the solution much simpler.

Linearity

The transform of a linear combination of signals is the same linear combination of their individual transforms. If

$$x_1(t) \leftrightarrow X_1(s) \quad \text{and} \quad x_2(t) \leftrightarrow X_2(s)$$

then

$$ax_1(t) + bx_2(t) \leftrightarrow aX_1(s) + bX_2(s)$$

where a and b are arbitrary constants. This property is the basis for transform analysis. Complex signals and their transforms can be easily obtained by decomposing them into a linear combination of simpler signals. Further, the inverse transform is easily obtained.

2.1.2   Time-Shifting

If $x(t)u(t) \leftrightarrow X(s)$, then

$$x(t - t_0)u(t - t_0) \leftrightarrow e^{-st_0}X(s), \quad t_0 \geq 0$$

Delaying a signal by t_0 seconds amounts to multiplying its transform with e^{-st_0}.
Consider the signal $x(t)u(t) = e^{-2t}u(t)$ and its shifted version $e^{-2(t-3)}u(t - 3)$.
The Laplace transforms of the two signals are, respectively, $\frac{1}{s+2}$ and $\frac{e^{-3s}}{s+2}$.

2.1.3   Frequency-Shifting

If $x(t)u(t) \leftrightarrow X(s)$, then

$$e^{s_0t}x(t)u(t) \leftrightarrow X(s - s_0)$$

Multiplying the signal $x(t)$ by the exponential e^{s_0t} amounts to changing the complex frequency of its spectral components by s_0. Therefore, the spectrum $X(s)$ is shifted in the s-plane by the amount s_0.
For example,

$$\sin(\omega_0 t)u(t) \leftrightarrow \frac{\omega_0}{s^2 + \omega_0^2}, \quad \text{Re}(s) > 0$$

$$\cos(\omega_0 t)u(t) \leftrightarrow \frac{s}{s^2 + \omega_0^2}, \quad \text{Re}(s) > 0$$

Then,

$$e^{-at}\cos(\omega_0 t)\,u(t) \leftrightarrow \frac{s + a}{(s + a)^2 + \omega_0^2}, \quad \text{Re}(s) > -a$$

$$e^{-at}\sin(\omega_0 t)\,u(t) \leftrightarrow \frac{\omega_0}{(s + a)^2 + \omega_0^2}, \quad \text{Re}(s) > -a$$

2.1.4   Time-Differentiation

This property expresses the transform of the time derivative of a signal in terms of its transform. If $x(t) \leftrightarrow X(s)$, then

$$\frac{d\,x(t)}{dt} \leftrightarrow sX(s) - x(0^-)$$

where $x(0^-)$ the initial value of the signal at $t = 0^-$. As the signal is expressed in terms of exponentials of the form $X(s)e^{st}$, the transform of the derivative is $sX(s)$, as given by the first term. The second term is the derivative at $t = 0$. This property makes the analysis of systems with initial conditions easier.

This property can be extended, by repeated application, to find the transform of higher-order derivatives. For example,

$$\frac{d}{dt}\left(\frac{dx(t)}{dt}\right) = \frac{d^2x(t)}{dt^2}$$

$$\leftrightarrow s(sX(s) - x(0^-)) - \frac{dx(t)}{dt}|_{t=0^-} = s^2X(s) - sx(0^-) - \frac{dx(t)}{dt}|_{t=0^-}$$

The entity

$$\frac{dx(t)}{dt}|_{t=0^-}$$

is the value of the derivative of $x(t)$ at $t = 0^-$.

Consider the transform pair

$$\cos(\omega_0 t)u(t) \leftrightarrow \frac{s}{s^2 + \omega_0^2}$$

Let us find the transform of $\sin(\omega_0 t)u(t)$ from that of $\cos(\omega_0 t)u(t)$ using the time-differentiating property. The derivative of $\cos(\omega_0 t)u(t)$ in the time domain is $-\omega_0 \sin(\omega_0 t)u(t)$. The transform of $\sin(\omega_0 t)u(t)$ is

$$-\frac{1}{\omega_0}\left(s\frac{s}{s^2 + \omega_0^2} - 1\right) = \frac{\omega_0}{s^2 + \omega_0^2}$$

Consider the input–output relationship of an inductor

$$v(t) = L\frac{di(t)}{dt}$$

where $v(t)$ is the voltage across the inductor and $i(t)$ is the current flowing through it. The value of the inductor is L henries. Let the initial value of current in the inductor be $i(0^-)$. From the time-differentiating property, we get the value of the voltage across the inductor as, in the frequency-domain,

$$V(s) = L(sI(s) - i(0^-))$$

The linearity and time-differentiation properties reduce the differential equation characterizing a system into algebraic equations, making the analysis simpler. In addition, transform methods bring out the salient properties of signals and systems.

2.1.5 Integration

If $x(t) \leftrightarrow X(s)$, then

$$\int_{0^-}^{t} x(\tau)\, d\tau \leftrightarrow \frac{1}{s} X(s)$$

As the integration can be considered as the inverse of the derivative property for a signal with zero DC component, the variable s appears in the denominator in the transform of the integral of the signal. Since the unit-ramp signal is the integral of the unit-step, $tu(t) \leftrightarrow 1/s^2$. Since the unit-parabolic signal is the integral of the unit-ramp, $0.5t^2u(t) \leftrightarrow 1/s^3$.

Of particular interest to system analysis is that this property enables easier analysis of components with initial conditions, such as a capacitor with some initial voltage. The input–output relationship of a capacitor is

$$v(t) = \frac{1}{C} \int_{0^-}^{t} i(\tau)d\tau + v(0^-)$$

where $v(0^-)$ is the initial value of voltage across the capacitor, $v(t)$ is the voltage across the capacitor and $i(t)$ is the current flowing through it. The value of the capacitor is C farads. From the time-integration property, we get the value of the voltage across capacitor as, in the frequency-domain,

$$V(s) = \frac{I(s)}{sC} + \frac{v(0^-)}{s}$$

2.1.6 Time-Scaling

If $x(t)u(t) \leftrightarrow X(s)$, then

$$x(at)u(at) \leftrightarrow \frac{1}{a} X\left(\frac{s}{a}\right), \quad a > 0$$

From the definition, the Laplace transform of $x(at)u(at)$ is

$$\int_{0^-}^{\infty} x(at)u(at)e^{-st} dt$$

Substituting $at = \tau$, we get $t = \frac{\tau}{a}$ and $dt = \frac{d\tau}{a}$. For $a > 0$, $u(at) = u(t)$. Incorporating the changes, we get

$$\frac{1}{a}\int_{0^-}^{\infty} x(\tau)e^{-\frac{s}{a}\tau}d\tau = \frac{1}{a}X\left(\frac{s}{a}\right)$$

Frequency and time are inversely affected by the scale factor a. Compression (expansion) of a signal in the time domain, by changing t to at, results in the expansion (compression) of its spectrum with the change s to $\frac{s}{a}$, in addition to scaling by $\frac{1}{a}$ (to take into account of the change in energy or power).

For example, consider the transform pair

$$\cos(4t)u(t) \leftrightarrow \frac{s}{s^2+4^2}$$

Let $a = 2$. Then,

$$\cos(8t)u(t) = \leftrightarrow 0.5\frac{s/2}{(s/2)^2+4^2} = \frac{s}{s^2+8^2}$$

The frequency gets multiplied by $a = 2$.

2.1.7 Convolution in Time

If $x(t)u(t) \leftrightarrow X(s)$ and $h(t)u(t) \leftrightarrow H(s)$, then

$$y(t) = x(t)u(t) * h(t)u(t) = \int_0^{\infty} x(\tau)h(t-\tau)d\tau \leftrightarrow Y(s) = X(s)H(s)$$

From the Laplace transform definition,

$$Y(s) = \int_0^{\infty}\int_0^{t} x(\tau)h(t-\tau)e^{-st}d\tau dt$$

Changing the order of integration, we get

$$Y(s) = \int_0^{\infty}\int_{\tau}^{\infty} x(\tau)h(t-\tau)e^{-st}dtd\tau$$

$$= \int_0^{\infty} x(\tau)e^{-\tau s}\left(\int_{\tau}^{\infty} h(t-\tau)e^{-s(t-\tau)}dt\right)d\tau$$

Replacing $t - \tau$ by v in the inner integral, we get

$$Y(s) = \int_0^{\infty} x(\tau)e^{-s\tau}\left(\int_0^{\infty} h(v)e^{-sv}dv\right)d\tau = X(s)H(s)$$

Consider the convolution of $e^{-t}u(t)$ and $e^{-2t}u(t)$. The inverse of the product of their transforms,

$$\frac{1}{(s+1)(s+2)} = \left(\frac{1}{(s+1)} - \frac{1}{(s+2)}\right),$$

is the convolution output $(e^{-t} - e^{-2t})u(t)$.

2.1.8   Multiplication by t

If $x(t)u(t) \leftrightarrow X(s)$, then

$$tx(t)u(t) \leftrightarrow -\frac{dX(s)}{ds}$$

Differentiating the defining expression for $-X(s)$ with respect to s, we get

$$-\frac{dX(s)}{ds} = -\frac{d}{ds}\left(\int_{0^-}^{\infty} x(t)u(t)e^{-st}dt\right) = \int_{0^-}^{\infty} tx(t)e^{-st}dt$$

In general,

$$t^n x(t)u(t) \leftrightarrow (-1)^n \frac{d^n X(s)}{ds^n}, \quad n = 0, 1, 2, \ldots$$

For example,

$$\frac{t}{2}u(t) \leftrightarrow \frac{1}{2s^2} \quad \text{and} \quad \frac{t^2}{2}u(t) \leftrightarrow -\frac{d(\frac{1}{2s^2})}{ds} = \frac{1}{s^3}$$

2.1.9   Initial Value

The initial and final values of a function can be easily found using the initial and final value properties of the transform. These values can be used to check the transforms of the signals. If $x(t) \leftrightarrow X(s)$ and the degree of the numerator polynomial of $X(s)$ is less than that of the denominator polynomial, then

$$x(0^+) = \lim_{s \to \infty} sX(s)$$

Examples are cosine and sine functions. Using this property, we can verify that the initial values are, respectively, 1 and 0.

2.1.10 Final Value

If $x(t) \leftrightarrow X(s)$ and the ROC of $sX(s)$ includes the $j\omega$ axis, then

$$\lim_{t \to \infty} x(t) = \lim_{s \to 0} sX(s)$$

We get the final value as 1, when this property is applied to the unit-step signal.

A list of Laplace transform pairs is shown in Table 2.1. A list of Laplace transform properties is shown in Table 2.2.

Table 2.1 Laplace transform pairs

$x(t)$	$X(s)$	ROC
$\delta(t)$	1	All s
$u(t)$	$\frac{1}{s}$	$\mathrm{Re}(s) > 0$
$t^n u(t)$, $n = 0, 1, 2, \ldots$	$\frac{n!}{s^{n+1}}$	$\mathrm{Re}(s) > 0$
$e^{-at} u(t)$	$\frac{1}{s+a}$	$\mathrm{Re}(s) > -a$
$t^n e^{-at} u(t)$, $n = 0, 1, 2, \ldots$	$\frac{n!}{(s+a)^{n+1}}$	$\mathrm{Re}(s) > -a$
$\cos(\omega_0 t) u(t)$	$\frac{s}{s^2+\omega_0^2}$	$\mathrm{Re}(s) > 0$
$\sin(\omega_0 t) u(t)$	$\frac{\omega_0}{s^2+\omega_0^2}$	$\mathrm{Re}(s) > 0$
$e^{-at} \cos(\omega_0 t) u(t)$	$\frac{s+a}{(s+a)^2+\omega_0^2}$	$\mathrm{Re}(s) > -a$
$e^{-at} \sin(\omega_0 t) u(t)$	$\frac{\omega_0}{(s+a)^2+\omega_0^2}$	$\mathrm{Re}(s) > -a$
$t \cos(\omega_0 t) u(t)$	$\frac{s^2-\omega_0^2}{(s^2+\omega_0^2)^2}$	$\mathrm{Re}(s) > 0$
$t \sin(\omega_0 t) u(t)$	$\frac{2\omega_0 s}{(s^2+\omega_0^2)^2}$	$\mathrm{Re}(s) > 0$

Table 2.2 Laplace transform properties

Property	$x(t)u(t), h(t)u(t)$	$X(s), H(s)$
Linearity	$ax(t) + bh(t)$	$aX(s) + bH(s)$
Time-shifting	$x(t - t_0)u(t - t_0)$, $t_0 \geq 0$	$X(s)e^{-st_0}$
Frequency-shifting	$x(t)u(t)e^{s_0 t}$	$X(s - s_0)$
Time-convolution	$x(t) * h(t)$	$X(s)H(s)$
Time-scaling	$x(at)$, $a > 0$ and real	$\frac{1}{a}X(\frac{s}{a})$
Time-differentiation	$\frac{dx(t)}{dt}$	$sX(s) - x(0^-)$
Time-differentiation	$\frac{d^2 x(t)}{dt^2}$	$s^2 X(s) - sx(0^-) - \frac{dx(t)}{dt}\vert_{t=0^-}$
Time-integration	$\int_{0^-}^{t} x(\tau)d\tau$	$\frac{X(s)}{s}$
Time-integration	$\int_{-\infty}^{t} x(\tau)d\tau$	$\frac{X(s)}{s} + \frac{1}{s}\int_{-\infty}^{0^-} x(\tau)d\tau$
Frequency-differentiation	$tx(t)u(t)$	$-\frac{dX(s)}{ds}$
Frequency-differentiation	$t^n x(t)u(t)$, $n = 0, 1, 2, \ldots$	$(-1)^n \frac{d^n X(s)}{ds^n}$
Initial value	$x(0^+)$	$\lim_{s \to \infty} sX(s)$, if $X(s)$ is strictly proper
Final value	$\lim_{t \to \infty} x(t)$	$\lim_{s \to 0} sX(s)$, (ROC of $sX(s)$ includes the $j\omega$ axis)

2.2 Laplace Transform Solution of Differential Equations

As we indicated initially, systems are characterized, usually, by differential equations, which are difficult to solve and, hence, to find the system response. Now, we want to find the response of linear systems using the Laplace transform with algebraic manipulations alone, rather than solving differential equations. Finding the system response is a major application of the Laplace transform.

2.2.1 The Transfer Function

The following steps are used to find the transfer (system) function of a linear time-invariant system.

1. Finding the differential equation of the system using the laws governing its components and the configuration of the system
2. Assuming that the system is relaxed (all initial values are zero), take the Laplace transform of the differential equation
3. Specify the system input and output variables
4. The ratio of the output and the input, in the Laplace domain, is the transfer function of the system

The input and output of a system may be of different types. For example, the input to an electric motor is electrical and the output is mechanical rotation. The total output of linear system is the sum of its zero-input and zero-state components. The zero-input component is the response with the input zero and, therefore, due to initial conditions alone. The zero-state component is the response due to the input alone with the values of all the initial conditions zero. The two responses can be computed independently. As we assumed the system is relaxed in defining the transfer function, the system response obtained using it is the zero-state response. Compared with other system models, the transfer function provides a closed-form expression for the response of linear systems to input signals.

The order of a differential equation, a fixed-integer, is the highest power of the derivative of its output terms. While the order of a differential equation is N in general, for simplicity, consider the second-order differential equation of a causal LTI continuous system relating the input $x(t)$ and the output $y(t)$,

$$\frac{d^2y}{dt^2} + a_1\frac{dy}{dt} + a_0y = b_2\frac{d^2x}{dt^2} + b_1\frac{dx}{dt} + b_0x$$

where $a_2, a_1, a_0, b_2, b_1, b_0$ are constants. They are functions of the system components alone and, in no way, depends on the input and output signals. Taking the Laplace transform of both sides, we get, assuming initial conditions are all zero,

$$(s^2 + a_1 s + a_0)Y(s) = (b_2 s^2 + b_1 s + b_0)X(s)$$

The transfer function $H(s)$, which is the ratio of the transforms of the output and the input signals with the initial conditions zero, is obtained as

$$H(s) = \frac{Y(s)}{X(s)} = \frac{b_2 s^2 + b_1 s + b_0}{s^2 + a_1 s + a_0} = \frac{\sum_{l=0}^{2} b_l s^l}{s^2 + \sum_{l=0}^{1} a_l s^l}$$

In general,

$$H(s) = \frac{Y(s)}{X(s)} = \frac{b_M s^M + b_{M-1} s^{M-1} + \cdots + b_1 s + b_0}{s^N + a_{N-1} s^{N-1} + \cdots + a_1 s + a_0}$$

If the input to the system is the unit-impulse signal, then its transform is one and $H(s) = Y(s)$. That is, the transform of the impulse response is the transfer function of the system. For stable systems, the frequency response $H(j\omega)$ is obtained from $H(s)$ by replacing s by $j\omega$. $H(s)$ is called the transfer function, since the input is transferred to output by multiplication with it. That is, $Y(s) = H(s)X(s)$. $Y(s)$ is the system response, in the Laplace domain, due to input $X(s)$. As the time-domain response is required most of the time, we can get it by finding the inverse Laplace transform of $Y(s)$.

2.2.2   Transfer Function of Feedback Systems

Consider the two systems connected in a feedback configuration, shown in Fig. 2.1. The feedback signal $R(s)$ can be expressed as $R(s) = F(s)Y(s)$, where $F(s)$ is the **feedback transfer function** of the system and $Y(s)$ is the output. Now, the error signal $E(s)$ is

$$E(s) = X(s) - R(s) = X(s) - F(s)Y(s)$$

The output $Y(s)$ is expressed as

$$Y(s) = G(s)E(s) = G(s)(X(s) - F(s)Y(s))$$

where $G(s)$ is the **feedforward or open-loop transfer function** of the system. The **closed-loop transfer function** of the feedback system is given as

$$H(s) = \frac{Y(s)}{X(s)} = \frac{G(s)}{1 + G(s)F(s)}$$

Fig. 2.1 Two systems
connected in a feedback
configuration

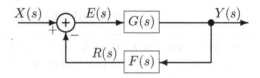

A system with this transfer function is equivalent to the system shown in Fig. 2.1
and it is more appropriate to use $H(s)$ to find the system response to any input.

If $G(s)$ is very large, the transfer function of the feedback system approximates
to the inverse of the feedback transfer function of the system.

$$H(s) = \frac{Y(s)}{X(s)} \approx \frac{1}{F(s)}$$

The **loop transfer function** is $L(s) = G(s)F(s)$. That is, the product of all
the terms around the loop, not including the -1 at the summing junction. In unity
feedback systems, obviously, the open-loop and loop transfer functions are the same.
We often use unity feedback systems in the analysis and design of control systems.
In terms of $L(s)$, the closed-loop transfer function is

$$H(s) = \frac{G(s)}{1 + L(s)}$$

and the characteristic equation is $1+L(s) = 0$. It is possible to get much information
about the behavior of the feedback system from $L(s)$ alone.

As the Laplace transform is used, this transfer function is the ratio of the Laplace
transforms of the output and input signals with all the initial conditions set to 0 (the
system is initially relaxed). The equation, obtained by equating the denominator
polynomial of a transfer function to zero, is the **characteristic equation**,

$$1 + G(s)F(s) = 0$$

as the roots of this equation determines the form of the system response.

Typically, the feedback configuration is used to design a system with the desired
response. The system with transfer function $G(s)$ may be unstable or its response
may not be acceptable. In order to make it to give the desired response with a
sufficient stability margin, we use the feedback configuration.

2.3   Finding the Inverse Laplace Transform

The inverse Laplace transform of $X(s)$ is defined as

$$x(t) = \frac{1}{2\pi j} \int_{\sigma-j\infty}^{\sigma+j\infty} X(s)e^{st}ds \quad \text{for } t \geq 0$$

where σ is any real value that lies in the ROC of $X(s)$. Note that the integral converges to the value zero for $t < 0$ and to the mid-point value at any discontinuity of $x(t)$. This equation is not often used for finding the inverse transform, as it requires integration in the complex plane.

2.3.1 Inverse Laplace Transform by Partial-Fraction Expansion

In the transform method of system analysis, we find the forward transform of signals, do the required processing in the transform domain, and find the inverse transform to get the time-domain version of the processed signal. Most of the Laplace transforms of practical interest are rational functions (a ratio of two polynomials in s). The denominator polynomial can be factored into a product of first- or second-order terms. This type of Laplace transforms can be expressed as the sum of partial fractions with each denominator forming a factor. The inverse Laplace transforms of the individual fractions can be easily found from a short table of transform pairs, such as those of $\delta(t)$, $u(t)$, $tu(t)$, $t^2u(t)$, $e^{-at}u(t)$, and $te^{-at}u(t)$, shown in Table 2.1. The sum of the individual inverses is the inverse of the given Laplace transform.

Two rational functions are added by converting them to a common denominator, add and then simplify. For example, the sum of the two rational functions is

$$X(s) = \frac{2}{(s+1)} + \frac{3}{(s+2)} = \frac{2(s+2)+3(s+1)}{(s+1)(s+2)} = \frac{(5s+7)}{(s^2+3s+2)}$$

Usually, we are given $X(s)$ to be inverted in the form on the right. The task is to find an equivalent expression like that on the left. The numerator polynomial of the rightmost expression is of order 1, whereas that of the denominator is of order 2. Partial fraction expansion of a rational function expresses it as a sum of appropriate fractions with the coefficient of each fraction to be found.

Example 2.3 Find the zero-state response of the system governed by the differential equation

$$\frac{dy}{dt} + 4y(t) = \frac{dx}{dt} + 2x(t)$$

with the input $x(t) = u(t)$, the unit-step function.

Solution The Laplace transforms of the terms of the differential equation are

$$x(t) \leftrightarrow \frac{1}{s}, \quad \frac{dx}{dt} \leftrightarrow 1 \quad y(t) \leftrightarrow Y(s), \quad \frac{dy}{dt} \leftrightarrow sY(s)$$

Substituting the corresponding transform for each term in the differential equation and solving for $Y(s)$, we get

$$Y(s) = \frac{s+2}{s(s+4)} = \frac{A}{s} + \frac{B}{(s+4)}$$

$$A = \frac{(s+2)}{(s+4)}\bigg|_{s=0} = 0.5, \qquad B = \frac{(s+2)}{s}\bigg|_{s=-4} = 0.5$$

$$Y(s) = \frac{0.5}{s} + \frac{0.5}{(s+4)}$$

Taking the inverse Laplace transform, we get the complete response.

$$y(t) = (0.5 + 0.5e^{-4t})u(t)$$

The steady-state response is $0.5u(t)$ and the transient response is $(0.5e^{-4t})u(t)$. Letting $t = 0$, $y(0) = 1$. Letting $t \to \infty$, $y(\infty) = 0.5$. From the initial and final value properties also, we get

$$\lim_{s \to \infty} s\frac{0.5}{s} + s\frac{0.5}{(s+4)} = 1, \qquad \lim_{s \to 0} s\frac{0.5}{s} + s\frac{0.5}{(s+4)} = 0.5$$

While the problem is already solved, let us use the linearity property to get the solution. Taking advantage of linearity property, the problem can be split into two simpler problems

$$\frac{dy}{dt} + 4y(t) = 2x(t) \quad \text{and} \quad \frac{dy}{dt} + 4y(t) = \frac{dx}{dt}$$

with the input $x(t) = u(t)$, the unit-step function.
    The solution to $u(t)$ alone is

$$(0.25 - 0.25e^{-4t})u(t)$$

Since there is a factor 2, the solution to the first part is

$$(0.5 - 0.5e^{-4t})u(t)$$

The derivative is

$$(2e^{-4t})u(t)$$

Verify the solution by substituting it back into the differential equation.

$$(2e^{-4t})u(t) + 4(0.5 - 0.5e^{-4t}) = 2$$

The solution satisfies the differential equation. Consider the second part of the problem. The input is $\delta(t)$. Since $\delta(t)$ is the derivative of $u(t)$, the solution is the derivative of

$$(0.25 - 0.25e^{-4t})u(t)$$

which is

$$(e^{-4t})u(t)$$

The total solution is the sum of the solutions of the two parts.

$$(0.5 - 0.5e^{-4t})u(t) + (e^{-4t})u(t) = (0.5 + 0.5e^{-4t})u(t)$$

as obtained earlier.

Again, due to linearity, the impulse response can be found by finding the step response and differentiating it. This solution satisfies the derivative of

$$\frac{dy}{dt} + 4y(t) = u(t)$$

which is

$$\frac{d^2y}{dt^2} + 4\frac{dy}{dt} = \delta(t)$$

Since the impulse function is defined by its unit area, we have to integrate it to get a numerical value. The value $y(0^+)$ is obtained by integrating the differential equation from $t = 0^-$ to $t = 0^+$.

$$\int_{0^-}^{0^+} \frac{d^2y(t)}{dt^2}dt + \int_{0^-}^{0^+} 4\frac{dy}{dt}dt = \int_{0^-}^{0^+} \delta(t)dt$$

The right-hand side is equal to one. The first term on the left-hand side reduces to $y(0^+)$ as $y(0^-) = 0$, as the system is initially relaxed. The second term on the left side reduces to zero, as that function must be continuous in the vicinity of $t = 0$. Therefore, the equation reduces to $y(0^+) = 1$. The solution to the second part of the problem is also verified. ∎

Rational Function with the Same Order of Numerator and Denominator

Partial-fraction expansion is applicable, only if the degree of the numerator polynomial is less than that of the denominator. In practical system analysis, we encounter only rational functions with the degree of the numerator polynomial less than or equal to that of the denominator. In the case the degrees are equal, we divide the numerator polynomial by the denominator polynomial once to get a constant plus a proper function.

Example 2.4 Find the inverse of

$$X(s) = \frac{s+2}{s+4}$$

Solution

$$X(s) = \frac{s+2}{(s+4)} = 1 - \frac{2}{(s+4)}$$

Taking the inverse Laplace transform, we get the complete response.

$$x(t) = \delta(t) - 2e^{-4t}u(t)$$

Example 2.5 Find the zero-state response of the system governed by the differential equation

$$\frac{d^2y}{dt^2} + \frac{dy}{dt} + 2y = x,$$

using the Laplace transform, with the input $x(t) = e^{-2t}u(t)$.

Solution The Laplace transforms of the terms of the differential equation are

$$x(t) \leftrightarrow \frac{1}{(s+2)}, \quad y(t) \leftrightarrow Y(s), \quad \frac{dy}{dt} \leftrightarrow sY(s), \quad \frac{d^2y}{dt^2} \leftrightarrow s^2Y(s)$$

Substituting the corresponding transform for each term in the differential equation and solving for $Y(s)$, we get

$$Y(s) = \frac{1}{(s^2+s+2)(s+2)} = \frac{0.25}{(s+2)} + \frac{0.1890\angle - 2.2935}{(s+0.5000 - j1.3229)}$$

$$+ \frac{0.1890\angle + 2.2935}{(s+0.5000 + j1.3229)}$$

$$= \frac{0.25}{(s+2)} + \frac{-0.1250 - j0.1417}{(s+0.5000 - j1.3229)} + \frac{-0.1250 + j0.1417}{(s+0.5000 + j1.3229)}$$

Suppose the two complex-conjugate roots are

$$-a + jb \quad \text{and} \quad -a - jb$$

and the corresponding partial-fraction coefficients are

$$Re^{j\theta} \quad \text{and} \quad Re^{-j\theta}$$

Then, the corresponding inverse transformation is

$$2Re^{-at}\cos(bt + \theta)$$

Taking the inverse Laplace transform, we get the complete response.

$$y(t) = (0.25e^{-2t} + 0.3780e^{-0.5t}\cos(1.3229t - 2.2935))u(t)$$

Multiple-Order Poles

Each repeated linear factor $(s + a)^m$ contributes a sum of the form

$$\frac{A_m}{(s+a)^m} + \frac{A_{m-1}}{(s+a)^{m-1}} + \cdots + \frac{A_1}{(s+a)}$$

Example 2.6 Find the zero-state response of the system governed by the differential equation

$$\frac{d^3y}{dt^3} + 6\frac{d^2y}{dt^2} + 12\frac{dy}{dt} + 8y = \frac{d^2x}{dt^2} + \frac{dx}{dt} + x,$$

using the Laplace transform, with the input $x(t) = e^{-t}u(t)$.

Solution The Laplace transforms of the terms of the differential equation are

$$x(t) \leftrightarrow \frac{1}{(s+1)}, \quad \frac{dx}{dt} \leftrightarrow \frac{s}{(s+1)}, \quad \frac{d^2x}{dt^2} \leftrightarrow \frac{s^2}{(s+1)}$$

$$y(t) \leftrightarrow Y(s), \quad \frac{dy}{dt} \leftrightarrow sY(s), \quad \frac{d^2y}{dt^2} \leftrightarrow s^2Y(s)$$

Substituting the corresponding transform for each term in the differential equation and solving for $Y(s)$, we get

$$Y(s) = \frac{s^2 + s + 1}{(s^3 + 6s^2 + 12s + 8)(s + 1)} = \frac{A}{(s+2)} + \frac{B}{(s+2)^2} + \frac{C}{(s+2)^3} + \frac{D}{(s+1)}$$

$$D = \frac{s^2 + s + 1}{(s^3 + 6s^2 + 12s + 8)}\bigg|_{s=-1} = 1$$

$$C = \frac{s^2 + s + 1}{(s + 1)}\bigg|_{s=-2} = -3$$

With C and D known, to find A, one method is to multiply both sides by s and let $s \to \infty$. That is,

$$\lim_{s \to \infty} s \frac{s^2 + s + 1}{(s^3 + 6s^2 + 12s + 8)(s + 1)} = \lim_{s \to \infty} \frac{As}{(s + 2)} + \frac{Bs}{(s + 2)^2} + \frac{Cs}{(s + 2)^3} + \frac{Ds}{(s + 1)}$$

We get $0 = 1 + A$ or $A = -1$. To find B, we replace s by a value other than the roots. Let $s = 0$ and we get

$$\frac{0 + 0 + 1}{(0 + 0 + 0 + 8)(0 + 1)} = \frac{1}{(0 + 2)} + \frac{B}{(0 + 2)^2} + \frac{-3}{(0 + 2)^3} + \frac{1}{(0 + 1)} \text{ or } B = 0$$

Now,

$$Y(s) = \frac{s^2 + s + 1}{(s^3 + 6s^2 + 12s + 8)(s + 1)} = \frac{-1}{(s + 2)} + \frac{0}{(s + 2)^2} + \frac{-3}{(s + 2)^3} + \frac{1}{(s + 1)}$$

Taking the inverse Laplace transform, we get the complete response.

$$y(t) = (e^{-t} - e^{-2t} - 1.5t^2 e^{-2t})u(t)$$

Example 2.7 Find the inverse of

$$X(s) = \frac{1}{(s^2 + 1)^2}$$

Solution First, let us find the inverse using properties.

$$\sin(t) \leftrightarrow \frac{1}{(s^2 + 1)}$$

Applying frequency-differentiating property, we get

$$t \sin(t) \leftrightarrow \frac{2s}{(s^2 + 1)^2}$$

Now,

$$\frac{1}{(s^2 + 1)^2} = \left(\frac{2s}{(s^2 + 1)^2} \right) \left(\frac{1}{2s} \right)$$

Using the convolution property, the inverse is

$$\frac{1}{2} * t \sin(t) = \frac{1}{2} \int_0^t \tau \sin(\tau) d\tau = \frac{1}{2} (\sin(t) - t \cos(t)) u(t)$$

By partial fraction, we get

$$X(s) = \frac{1}{(s^2 + 1)^2} = \frac{j0.25}{s + j} + \frac{-j0.25}{s - j} + \frac{-0.25}{(s + j)^2} + \frac{-0.25}{(s - j)^2}$$

$$= \frac{0.5}{(s^2 + 1)} - \frac{0.5(s^2 - 1)}{(s^2 + 1)^2}$$

Taking the inverse, we get the same result.

2.4   Characterization of a System by Its Poles and Zeros and System Stability

The numerator and denominator polynomials of the transfer function can be factored to get

$$H(s) = K \frac{(s - z_1)(s - z_2) \cdots (s - z_M)}{(s - p_1)(s - p_2) \cdots (s - p_N)} = K \frac{\prod_{l=1}^{M}(s - z_l)}{\prod_{l=1}^{N}(s - p_l)}$$

where K is a constant. The numerator evaluates to zero, when s takes on the values $z_1, z_2, \ldots, z_M$. That is, $H(s)$ itself evaluates to zero at these values. Therefore, the values $z_1, z_2, \ldots, z_M$ are called the **zeros** of $H(s)$ and represent frequencies in the s-plane. Similarly, the denominator evaluates to zero, when s takes on the values $p_1, p_2, \ldots, p_N$. That is, $H(s)$ tends to infinity at these values. Therefore, the values $p_1, p_2, \ldots, p_N$ are called the **poles** of $H(s)$. As the coefficients of the polynomials of $H(s)$ are real for practical systems, the zeros and poles are real-valued or they always occur as complex-conjugate pairs. The pole-zero representation of the transfer function $H(s)$ of a system gives a clear description of its characteristics, such as speed of response, frequency selectivity, and stability. Poles and zeros, along with a constant, is a complete description of a system.

The zero-input response of a system depends solely on the locations of its poles. A system is considered stable if its zero-input response due to finite initial conditions converges, marginally stable if its zero-input response tends to a constant value or oscillates with a constant amplitude, and unstable if its zero-input response diverges. Commonly used marginally stable systems are oscillators, which produce a bounded zero-input response.

Figure 2.2 shows pole locations of some transfer functions and the corresponding impulse responses. The response corresponding to each pole p of a system is of the form e^{at}, where a is the location of the pole in the s-plane. The poles corresponding to step and sine functions are shown in (a). As the real part is zero (located on the imaginary axis), responses remain bounded (oscillating with a constant amplitude or a constant value). The poles corresponding to decaying sinusoid and decaying real exponential functions are shown in (d). As the real part is negative (located

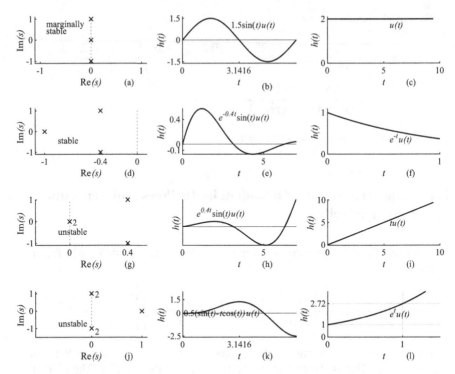

Fig. 2.2 The poles of some transfer functions $H(s)$ and the corresponding impulse responses $h(t)$. The imaginary axis is shown by a dotted line. (**a**) $H(s) = \frac{1.5}{s^2+1} = \frac{1.5}{(s+j)(s-j)}$ and $H(s) = \frac{2}{s}$; (**b**) $h(t) = 1.5\sin(t)u(t)$ and (**c**) $h(t) = 2u(t)$; (**d**) $H(s) = \frac{1}{(s+0.4)^2+1} = \frac{1}{(s+0.4+j)(s+0.4-j)}$ and $H(s) = \frac{1}{s+1}$; (**e**) $h(t) = e^{-0.4t}\sin(t)u(t)$ and (**f**) $h(t) = e^{-t}u(t)$; (**g**) $H(s) = \frac{1}{(s-0.4)^2+1} = \frac{1}{(s-0.4+j)(s-0.4-j)}$ and $H(s) = \frac{1}{s^2}$; (**h**) $h(t) = e^{0.4t}\sin(t)u(t)$ and (**i**) $h(t) = tu(t)$; (**j**) $H(s) = \frac{1}{(s^2+1)^2} = \frac{1}{(s+j)^2(s-j)^2}$ and $H(s) = \frac{1}{s-1}$; (**k**) $h(t) = 0.5(\sin(t) - t\cos(t))u(t)$ and (**l**) $h(t) = e^t u(t)$

on the left of the imaginary axis), responses decay, as shown in (e) and (f). The poles corresponding to growing sinusoid and ramp functions are shown in (g). As the real part is positive (located on the right of the imaginary axis) or a double pole on the imaginary axis, responses grow unbounded, as shown in (h) and (i). The poles corresponding to growing sinusoid and growing real exponential functions are shown in (j). As the real part is positive (located on the right of the imaginary axis) or a double pole on the imaginary axis, responses grow unbounded, as shown in (k) and (l).

Therefore, we conclude that, in terms of the locations of the poles of a system:

- All the poles, of any order, of a stable system must lie to the left of the imaginary axis of the s-plane. That is, the ROC of $H(s)$ must include the imaginary axis.
- Any pole lying to the right of the imaginary axis or any pole of order more than one lying on the imaginary axis makes a system unstable.

- A system is marginally stable if it has no poles to the right of the imaginary axis and has poles of order one on the imaginary axis.

An alternate stability condition is that bounded input produces bounded output. If all the poles of a system lie to the left of the imaginary axis of the s-plane, the bounded-input bounded-output stability condition is satisfied. However, the converse is not necessarily true, since the impulse response is an external description of a system and may not include all its poles. The bounded-input bounded-output stability condition is not satisfied by a marginally stable system.

2.5   Routh–Hurwitz Stability Criterion

Stability of systems is very important for practical systems. One reason is that the response of unstable systems will be unbounded even for bounded-input signals. Further, the response could burn out the system and lead to other undesirable effects. The characteristic equation is of the form (for example, that of a 6-th order system)

$$a_6 s^6 + a_5 s^5 + a_4 s^4 + a_3 s^3 + a_2 s^2 + a_1 s + a_0 = 0$$

which is the denominator polynomial of the transfer function equated to zero. By factoring the characteristic equation, of course, we find the locations of the poles of the system. While there are other methods for detailed stability analysis, this method answers the question whether the system is stable or not (absolute stability) with less computational effort and also determines the range of the open-loop gain K for stability. It does not explicitly compute the roots and applicable only if all the roots are real-valued. It determines the number of roots in the right-half of the s-plane. In practice, it is useful when the characteristic equation has at least one unknown parameter. This criterion basically requires the evaluation of a number of 2×2 determinants.

Construction of the Routh–Hurwitz Array
The steps are:

1. Label the rows with the highest power of the characteristic equation down to s^0. With the highest power N, there are $N + 1$ rows
2. In the first row, list every other coefficient of the characteristic equation, starting with that of the highest power
3. In the second row, list every other coefficient of the characteristic equation, starting with that of the next highest power
4. The entries in the remaining rows are determined by computing the negative of the values of 2×2 determinants divided by the first entry of their second row. The 2×2 determinant to determine the first entry of the new row is formed by the first entries of the immediately above two rows forming the first column of the determinant and the second entries of the immediately above two rows forming

Table 2.3 The Routh–Hurwitz array for a 6th order system with arbitrary coefficients

s^6	a_6		a_4		a_2	a_0
s^5	a_5		a_3		a_1	
s^4	$b_1 = -\frac{1}{a_5}\begin{vmatrix} a_6 & a_4 \\ a_5 & a_3 \end{vmatrix}$		$b_2 = -\frac{1}{a_5}\begin{vmatrix} a_6 & a_2 \\ a_5 & a_1 \end{vmatrix}$		$b_3 = a_0$	
s^3	$c_1 = -\frac{1}{b_1}\begin{vmatrix} a_5 & a_3 \\ b_1 & b_2 \end{vmatrix}$		$c_2 = -\frac{1}{b_1}\begin{vmatrix} a_5 & a_1 \\ b_1 & b_3 \end{vmatrix}$			
s^2	$d_1 = -\frac{1}{c_1}\begin{vmatrix} b_1 & b_2 \\ c_1 & c_2 \end{vmatrix}$		$d_2 = b_3$			
s^1	$e_1 = -\frac{1}{d_1}\begin{vmatrix} c_1 & c_2 \\ d_1 & d_2 \end{vmatrix}$					
s^0	$f_1 = d_2$					

Table 2.4 The Routh–Hurwitz array

s^6	1	20	25	2
s^5	7	30	11	
s^4	$\frac{110}{7}$ 55	$\frac{164}{7}$ 82	2, 7	
s^3	$\frac{1076}{55}$ 269	$\frac{556}{55}$ 139		
s^2	$\frac{14413}{269}$ 2059	269		
s^1	$\frac{213840}{2059}$			
s^0	269			

the second column. In general, the nth entry is computed using the determinant formed by the first entries and the nth entries of the immediately above two rows

Table 2.3 shows the Routh–Hurwitz array for the characteristic equation

$$a_6 s^6 + a_5 s^5 + a_4 s^4 + a_3 s^3 + a_2 s^2 + a_1 s + a_0 = 0$$

Example 2.8 Determine the stability of the system whose characteristic equation is given by

$$s^6 + 7s^5 + 20s^4 + 30s^3 + 25s^2 + 11s + 2 = 0$$

The equation has no missing terms and all the coefficients are of the same sign. This condition is a necessity but, still, the sufficient condition for stability must be checked from the Routh–Hurwitz array. Following the procedure shown in Table 2.3, we get Routh array shown in Table 2.4. The test remains the same when the coefficients of any row are multiplied or divided by a positive number in order to simplify the computation. The Routh's stability criterion is that:

Table 2.5 The
Routh–Hurwitz array

s^4	1	10	28
s^3	1	38	
s^2	-28	28	
s^1	39		
s^0	28		

The number of sign changes of the entries in the first column of the Routh–Hurwitz array is the number of roots of the characteristic equation in the right-half of the s-plane. If that number is zero, the system is stable

For this example, since all the entries in the first column have the same sign, the system is stable. That is, all the poles of the characteristic equation lie in the left-half of the s-plane. The roots are

$$\{-2, \ -1.0015, \ -1.0005 \pm j0.0015, \ -0.9988 \pm j0.0009\}$$

For a stable second-order system, all the coefficients of the characteristic equation must be positive or all the coefficients must be negative. For a stable third-order system, with the characteristic equation

$$a_3 s^3 + a_2 s^2 + a_1 s + a_0 = 0$$

all the coefficients must be positive or negative and $a_2 a_1 > a_0 a_3$.

Example 2.9 Determine the stability of the system whose characteristic equation is given by

$$s^4 + s^3 + 10s^2 + 38s + 28 = 0$$

This equation satisfies all the necessary conditions. That is, all the coefficients exist and are positive. The Routh array is shown in Table 2.5. Since there are two sign changes, there are two roots located in the right-half of the s-plane and the system is unstable. The four roots are

$$\{1 + j3.6056, \quad 1 - j3.6056, \quad -2, \quad -1\}$$

Two special cases in forming the Routh–Hurwitz array occur only for unstable systems.

1. Zero only in the first column
2. Entire row of zeros

Example 2.10 Determine the stability of the system whose characteristic equation is given by

$$s^3 + 3s^2 + s + 3 = 0$$

Table 2.6 The
Routh–Hurwitz array

s^3	1	1
s^2	3	3
s^1	0	
s^0	?	

Table 2.7 The
Routh–Hurwitz array

s^3	1	1
s^2	3	4
s^1	−1	
s^0	4	

This equation satisfies all the necessary conditions. That is, all the coefficients exist and are positive. In this case, the product of the two middle coefficients is not greater than that of the last and first, $(3 \times 1 = 3) = (3 \times 1 = 3)$, and therefore, the system is unstable. The Routh array is shown in Table 2.6. The 3rd row, consisting of one entry, is zero, making it impossible to proceed further. The factored form of the characteristic equation is

$$s^3 + 3s^2 + s + 3 = (s^2 + 1)(s + 3) = 0$$

The factors of the type $(s^2 + 1)$, which is the second row in Table 2.6, cause premature termination of the formation of the Routh–Hurwitz array. The roots are $\pm j$. After removing this factor from the characteristic equation, we can form the array for the rest of the factor $(s + 3)$. The root is obviously -3. The three roots are

$$\{j1, \quad -j1, \quad -3\}$$

Example 2.11 Determine the stability of the system whose characteristic equation is given by

$$s^3 + 3s^2 + s + 4 = 0$$

This equation satisfies all the necessary conditions. That is, all the coefficients exist and are positive. In this case, the product of the two middle coefficients is less than that of the last and first, $(3 \times 1 = 3) < (4 \times 1 = 4)$, and, therefore, the system is unstable. The Routh array is shown in Table 2.7. There are two sign changes and, therefore, two poles lie in the right-half of the s-plane. The three roots are

$$\{0.0473 + j1.1359, \quad 0.0473 - j1.1359, \quad -3.0946\}$$

Example 2.12 Determine the stability of the system whose characteristic equation is given by

$$s^4 + s^3 + 3s^2 + 3s + 1 = 0$$

Table 2.8 The
Routh–Hurwitz array

s^4	1	3	1
s^3	1	3	
s^2	$\cancel{0}\ \epsilon$	1	
s^1	$\frac{3\epsilon-1}{\epsilon} \approx \frac{-1}{\epsilon}$		
s^0	1		

Table 2.9 The
Routh–Hurwitz array

s^3	3	1	
s^2	8	3	
s^1	-1		
s^0	3		

This equation satisfies all the necessary conditions. That is, all the coefficients exist
and are positive. The Routh array is shown in Table 2.8. For small $\epsilon > 0$, the
fourth entry in the first column is negative indicating two sign changes. For small
$\epsilon < 0$, the third entry in the first column is negative indicating two sign changes.
The conclusion is that the system is unstable. The four roots are

$$\{0.0520+j1.6579, \quad 0.0520-j1.6579, \quad -0.5520+j0.2423, \quad -0.5520-j0.2423\}$$

A slightly easier method for this type of problems is to find the reciprocal
polynomial by replacing s in the given polynomial by $1/s$. Then, we get, for this
example,

$$1 + s + 3s^2 + 3s^3 + s^4 = 0$$

The coefficients appear in reverse order. The Routh array is shown in Table 2.9. The
first column entries show two sign changes, as determined by using ϵ method.

Example 2.13 Determine the stability of the system whose characteristic equation
is given by

$$s^4 + 3s^3 + 3s^2 + 3s + 2 = 0$$

This equation satisfies all the necessary conditions. That is, all the coefficients exist
and are positive. The Routh array is shown in Table 2.10. The only term in the s^1
row is zero. Then, an auxiliary polynomial is formed from the coefficients of the
previous row. $s^2 + 1$, which indicates that there is a pair of conjugate poles on the
imaginary axis. The derivative of $s^2 + 1$, with respect to s, is $2s$. The coefficient 2
replaces the zero-valued coefficient in the s^1 row. Of course, as all the coefficients
of a row can be divided by a positive constant, we divide 2 by 2 to get the final
coefficient for the s^1 row as 1. As there is no change in sign of the first column
coefficients, the rest of the poles are on the left side of the s-plane. The four roots
are

$$\{-2, \quad -1, \quad j, \quad -j\}$$

Table 2.10 The Routh–Hurwitz array

s^4	1	3	2	
s^3	3	3		
s^2	2̶ 1	2̶ 1		$\leftarrow s^2+1$
s^1	0̶ 2̶ 1			$\leftarrow s$
s^0	1			

Table 2.11 The Routh–Hurwitz array

s^3	1	3
s^2	4	K
s^1	$\frac{12-K}{4}$	> 0
s^0	K	> 0

The Routh–Hurwitz array can be used to determine the range of one or two parameters, such as gain K, for the system to be stable.

Example 2.14 Determine the range of K for the stability of the system whose characteristic equation is given by

$$s^3 + 4s^2 + 3s + K = 0$$

The Routh array is shown in Table 2.11. For the system to be stable,

$$\frac{12-K}{4} > 0 \quad \text{and} \quad K > 0 \qquad \text{or} \quad 0 < K < 12$$

With $K = 12$, the characteristic equation is given by

$$s^3 + 4s^2 + 3s + 12 = 0$$

and the roots are

$$-4, \quad j1.7321, \quad -j1.7321$$

Two poles are on the imaginary axis of the s-plane. For $K < 12$, all the poles are in the left-half of the s-plane.

Example 2.15 Determine the stability range of K for the system whose characteristic equation is given by

$$s^3 + 4Ks^2 + 3(K+1)s + 4 = 0$$

The Routh array is shown in Table 2.12. For the system to be stable,

$$3K^2 + 3K - 1 > 0 \quad \text{and} \quad K > 0$$

Table 2.12 The
Routh–Hurwitz array

s^3	1	$3(K+1)$
s^2	$4K$	4
s^1	$\frac{3K(K+1)-1}{K}$	> 0
s^0	4	

The roots of the quadratic equation are $\{0.2638, -1.2638\}$. As K cannot be negative, the condition is $K > 0.2638$. With $K = 0.2638$, the characteristic equation is given by

$$s^3 + 1.0553s^2 + 3.7914s + 4 = 0$$

and the roots are

$$-1.0551, \quad j1.9471, \quad -j1.9471$$

Two poles are on the imaginary axis of the s-plane. For $K > 0.2638$, all the poles are in the left-half of the s-plane.

2.6 Summary

- Time-domain (with the independent variable time) signals are transformed into equivalent frequency-domain signals (with the independent variable frequency) in the transformed representation.
- In the transform representation of signals, we use the complex exponential signal as basis functions to reduce the differentiation operation into much simpler multiplication operation.
- The complex exponential is an exponential function with a complex exponent. Basically, the signal is decomposed into complex exponentials with infinite number of complex frequencies. The operation involved in obtaining the transform representation for continuous signals is finding integrals of products of the signal to be transformed with complex exponentials.
- The inverse process involves complex integrals. However, the inverse can be obtained using a list of transforms of a few basic signals for most practical purposes.
- The one-sided or unilateral Laplace transform $X(s)$ of the time-domain signal $x(t) = 0$, $t < 0$ is defined as

$$X(s) = \int_{0-}^{\infty} x(t)e^{-st}dt$$

where $s = (\sigma + j\omega)$.

- Instead of using the definition for all the problems, the use of properties of the transform makes the problems much simpler.
- If $x(t) \leftrightarrow X(s)$ and the degree of the numerator polynomial of $X(s)$ is less than that of the denominator polynomial, then

$$x(0^+) = \lim_{s \to \infty} sX(s)$$

- If $x(t) \leftrightarrow X(s)$ and the ROC of sX(s) includes the $j\omega$ axis, then

$$\lim_{t \to \infty} x(t) = \lim_{s \to 0} sX(s)$$

- The total output of a linear system is the sum of its zero-input and zero-state components. The zero-input component is the response with the input zero and, therefore, due to initial conditions alone. The zero-state component is the response due to the input alone with the values of all the initial conditions zero.
- The transfer function $H(s)$ is the ratio of the transforms of the output and the input signals with the initial conditions zero.
- The closed-loop transfer function of the feedback system is given as

$$H(s) = \frac{Y(s)}{X(s)} = \frac{G(s)}{1 + G(s)F(s)}$$

where $F(s)$ is the feedback transfer function and $G(s)$ is the feedforward or open-loop transfer function.
- Most of the Laplace transforms of practical interest are rational functions (a ratio of two polynomials in s). The denominator polynomial can be factored into a product of first- or second-order terms. This type of Laplace transforms can be expressed as the sum of partial fractions with each denominator forming a factor. The inverse Laplace transforms of the individual fractions can be easily found from a short table of transform pairs, such as those of $\delta(t)$, $u(t)$, $tu(t)$, $t^2 u(t)$, $e^{-at} u(t)$, and $te^{-at} u(t)$. The sum of the individual inverses is the inverse of the given Laplace transform.
- A system is considered stable if its zero-input response due to finite initial conditions converges, marginally stable if its zero-input response tends to a constant value or oscillates with a constant amplitude, and unstable if its zero-input response diverges.
- Routh–Hurwitz stability criterion answers the question whether the system is stable or not (absolute stability) with less computational effort and also determines the range of the open-loop gain K for stability. It does not explicitly compute the roots and applicable only if all the roots are real-valued. It determines the number of roots in the right-half of the s-plane.

Exercises

2.1 Find the Laplace transform of the function $x(t)$ using the time-shifting property and the transforms of $u(t)$ and $tu(t)$.

2.1.1 $x(t) = u(t - 3)$.

2.1.2 $x(t) = 2$, $0 \le t \le 2$ and $x(t) = 0$ otherwise.

*2.1.3 $x(t) = tu(t - 2)$

2.2 Find the Laplace transform of the function $x(t)$ using the frequency-shifting property.

2.2.1 $e^{-t} \cos(2t)u(t)$.

*2.2.2 $e^{-2t} \sin(3t)u(t)$.

2.3 Given the Laplace transform $X(s)$ of $x(t)$, find $x(at)$ and its transform using the scaling property. Find the locations of the poles and zeros of the two transforms.

2.3.1 $X(s) = \frac{s+4}{s^2+5s+6}$ and $a = 2$.

2.3.2 $X(s) = \frac{s-1}{s^2+3s+2}$ and $a = 0.5$.

*2.3.3 $X(s) = \frac{1}{s^2+1}$ and $a = 3$.

2.4 Find the Laplace transform of the function $x(t)$ using the multiplication by t property.

*2.4.1 $t^2 u(t)$.

2.4.2 $t \sin(2t)u(t)$.

2.5 Find the initial and final values of the function $x(t)$ corresponding to the transform $X(s)$, using the initial and final value properties.

2.5.1 $X(s) = \frac{s+3}{(s+2)}$.

* 2.5.2 $X(s) = \frac{1}{s+4}$.

2.5.3 $X(s) = \frac{s}{s^2+1}$.

2.5.4 $X(s) = \frac{s^2+2s+2}{s(s^2+5s+4)}$.

2.5.5 $X(s) = \frac{s+1}{s(s-3)}$.

2.6 Find the inverse Laplace transform of

$$X(s) = \frac{s+3}{(s+2)}$$

*2.7 Find the inverse Laplace transform of

$$X(s) = \frac{s^2 + 2s + 2}{s(s^2 + 5s + 4)}$$

2.8 Find the inverse Laplace transform of

$$X(s) = \frac{s+2}{(s^2 + 1)}$$

*2.9 Find the inverse Laplace transform of

$$X(s) = \frac{s}{(s^3 + 4s^2 + 5s + 2)}$$

2.10 Find the inverse Laplace transform of

$$X(s) = \frac{s+3}{(s^3 + s^2)}$$

2.11 Find the inverse Laplace transform of

$$X(s) = \frac{s + e^{-s}}{(s+2)(s+3)}$$

*2.12 Find the inverse Laplace transform of

$$X(s) = \frac{se^{-s}}{(s+1)(s+2)}$$

*2.13 Find the response of the relaxed linear system governed by the linear differential equation

$$\frac{d^2 y}{dt^2} + 3\frac{dy}{dt} + 2y = 5\frac{dx}{dt} + 7x$$

using the Laplace transform. The input $x(t)$ is the unit-impulse, $x(t) = \delta(t)$.

2.14 Find the response of the relaxed linear system governed by the linear differential equation

$$\frac{d^2 y}{dt^2} + 2\zeta\omega_n\frac{dy}{dt} + \omega_n^2 y = \omega_n^2 x(t)$$

using the Laplace transform. Plot the response for $\zeta = \{0.4, 0.5, 0.6, 0.7, 0.8\}$ and $\omega_n = 2\pi$. The input $x(t)$ is the unit-step, $x(t) = u(t)$.

Chapter 3
Mathematical Modeling of Electrical Systems

An electrical system is an interconnection of components such as resistors, inductors, and capacitors to produce, transfer, and utilize electrical power. The modeling of electrical systems is important in control systems, since control systems are widely used in electrical engineering applications. Furthermore, models of other type of systems are often converted to those of analogous electrical systems to take advantage of the familiar and well-developed circuit theory. For the analysis of any type of systems, a mathematical model has to be derived with some assumptions, so that the complexity of the model depends on the accuracy required.

3.1 Modeling of Electrical Circuits

Using the voltage–current relationships of various components and the basic laws of interconnection, the desired outputs are found when the inputs are specified. Examples are presented to recollect the basics of circuit analysis studied in an earlier course. If necessary, a circuit theory book, such as that given at the end of the book, should be used as a reference.

3.1.1 Circuit Analysis

Basic Elements in Electrical Circuits

Figures 3.1a–c show, respectively, the input–output relationship of a resistor of value R ohms, an inductor of value L henries, and a capacitor of value C farads, in the time domain. Table 3.1 shows the input–output relationship of a resistor of value R ohms, an inductor of value L henries and a capacitor of value C farads in time and frequency domains. Each electrical component has a voltage–current relationship, such as Ohm's law for a resistor. The constraints (a condition that a solution to

D. Sundararajan, *Control Systems*, https://doi.org/10.1007/978-3-030-98445-8_3

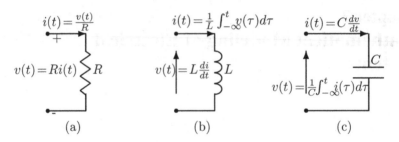

Fig. 3.1 Volt–ampere relationships of a resistor, an inductor, and a capacitor in the time domain

Table 3.1 Volt–ampere relationships of basic circuit elements in time and frequency domains

Element	Time domain	Frequency domain	Impedance
R	$v(t) = Ri(t)$	$V(s) = RI(s)$	$Z_R = R$
L	$v(t) = L\frac{di}{dt}$	$V(s) = L(sI(s) - i(0))$	$Z_L(s) = sL$
C	$v(t) = \frac{1}{C}\int_{-\infty}^{t} i(\tau)d\tau$	$V(s) = \frac{I(s)}{sC} + \frac{v(0)}{s}$	$Z_C(s) = \frac{1}{sC}$

a problem must satisfy) on voltages and currents in different parts of the system, called the circuit, must satisfy the well-known Kirchhoff's voltage and current laws.

Kirchhoff's voltage law (KVL): the algebraic sum of voltages across all branches around any closed loop of a circuit is zero at all instants of time.

Kirchhoff's current law (KCL): the algebraic sum of all currents at any node leaving through branches is zero at all instants of time.

Ohm's law: The current flowing through a resistor is directly proportional to the voltage applied across it and it is inversely proportional to its value at all instants of time.

The voltage across a capacitor is the time integral of the current flowing through it times the reciprocal of its value. The voltage across an inductor is the time derivative of the current flowing through it times its value.

3.1.2　Series Circuits

In a series connection, one, and only one, terminal of an element is connected to adjoining elements. Figure 3.2a shows the time-domain representation of a resistor, an inductance and a capacitor connected in cascade, called a series circuit. Figure 3.2b shows the frequency-domain representation of the same circuit. The advantage of this representation is that the analysis of the circuit becomes algebraic. A circuit is an interconnection of elements. The determination of currents and voltages at all parts of the circuit is the essence of circuit analysis. When impedances are connected in series, the voltage across them increases, with the same current

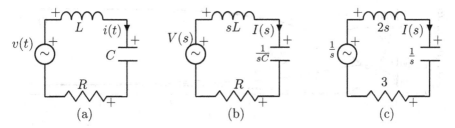

Fig. 3.2 Series circuit with a voltage source

flowing through them. It is similar to connecting hoses to make a longer hose. The combined impedance is the sum of all the impedances. That is, with N number of impedances connected in series, the equivalent impedance $Z_{eq}(s)$ of the series circuit is

$$Z_{eq}(s) = Z_1(s) + Z_2(s) + \cdots + Z_N(s)$$

The same current $I(s)$ passes through all the impedances. Therefore, the voltage $V(s)$ across the series connection is

$$V(s) = I(s)Z_1(s) + I(s)Z_2(s) + \cdots + I(s)Z_N(s) = I(s)Z_{eq}(s)$$

The equivalent impedance remains unchanged, irrespective of the order in which they are connected. Obviously, if all of them have the same value, then $Z_{eq}(s) = NZ(s)$. The source voltage applied across them gets divided in proportion to their individual values. The current through the series circuit is

$$I(s) = \frac{V(s)}{Z_{eq}(s)}$$

and the voltage across any impedance $Z_n(s)$ is

$$V_n(s) = \frac{V(s)}{Z_{eq}(s)} Z_n(s)$$

With just two impedances, $Z_1(s)$ and $Z_2(s)$, and $V(s)$, the applied voltage, in the series connection, we get

$$V_1(s) = \frac{V(s)}{Z_1(s) + Z_2(s)} Z_1(s) \quad \text{and} \quad V_2(s) = \frac{V(s)}{Z_1(s) + Z_2(s)} Z_2(s)$$

Consider the circuit shown in Fig. 3.2c. The circuit is energized by unit-step voltage source, $u(t) \leftrightarrow 1/s$. Given $R = 3$, $L = 2H$, and $C = 1F$, the corresponding impedances in the frequency domain are

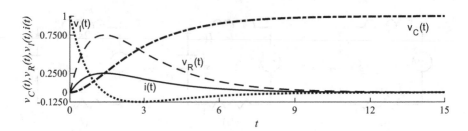

Fig. 3.3 The current through the series circuit

$$Z_R = 3, \quad Z_L(s) = 2s, \quad Z_C(s) = \frac{1}{s}$$

The impedances of the inductor and capacitor are a function of the frequency. The equivalent impedance is

$$Z_{eq}(s) = 3 + 2s + \frac{1}{s} = \frac{2s^2 + 3s + 1}{s}$$

The current through the series circuit, with $V(s) = 1/s$, is

$$I(s) = \frac{V(s)}{Z_{eq}(s)} = \frac{1}{2s^2 + 3s + 1} = -\frac{1}{s+1} + \frac{1}{s+0.5}$$

Taking the inverse Laplace transform, we get the current through the series circuit as

$$i(t) = (-e^{-t} + e^{-0.5t})u(t)$$

Using the initial and final value theorems of the Laplace transform, we get 0 and 0, which are also the same by getting them from $i(t)$. The voltage across the resistor is three times that of the current. Differentiating $i(t)$ and multiplying by 2, we get the voltage across as

$$v_L(t) = 2(e^{-t} - 0.5e^{-0.5t})$$

By integrating the current, we get the voltage across as

$$v_C(t) = (1 + e^{-t} - 2e^{-0.5t})$$

The voltages and current are shown in Fig. 3.3.

3.1.3 Parallel Circuits

In parallel connection, one terminal of all the elements is connected to one node and the other terminals are connected to another node. Therefore, the voltage across each element is equal. Figure 3.4 shows a resistor, an inductor, and a capacitor connected in parallel, called a parallel circuit. When elements are connected in parallel, while the voltage across all of them is the same, different currents flow through them, unless some or all of them are the same. It is similar to connecting hoses to make a wider hose. The length remains the same, but the flowing capacity increases. The combined admittance, the reciprocal of the impedance $Y(s) = 1/Z(s)$, is the sum of all the admittances. That is, with N number of elements connected in parallel, the equivalent admittance $Y_{eq}(s)$ of the parallel circuit is

$$Y_{eq}(s) = Y_1(s) + Y_2(s) + \cdots + Y_N(s) \quad \text{and} \quad Z_{eq}(s) = \frac{1}{Y_{eq}(s)}$$

where $Y_n(s) = 1/Z_n(s)$. The value of $Z_{eq}(s)$ will be smaller than the smallest of the impedances in the parallel connection, since the total current is more. The same voltage $V(s)$ is applied across all the elements. Therefore, the total current $I(s)$ flowing through the parallel connection is

$$I(s) = V(s)(Y_1(s) + Y_2(s) + \cdots + Y_N(s)) = V(s)Y_{eq}(s)$$

The equivalent admittance remains unchanged, irrespective of the order in which they are connected. Obviously, if all of them have the same value, then $Y_{eq}(s) = NY(s)$. The total current gets divided in proportion to their individual admittance values. The total current through the circuit is

$$I(s) = V(s)Y_{eq}(s)$$

and the current through any admittance $Y_n(s)$ is

$$I_n(s) = \frac{I(s)}{Y_{eq}(s)} Y_n(s)$$

With just two admittances, $Y_1(s)$ and $Y_2(s)$, and $I(s)$, the total current, we get

$$I_1(s) = \frac{I(s)}{Y_1(s) + Y_2(s)} Y_1(s) \quad \text{and} \quad I_2(s) = \frac{I(s)}{Y_1(s) + Y_2(s)} Y_2(s)$$

Consider the circuit shown in Fig. 3.4a. The circuit is energized by a causal current source of $\sin(t)u(t)$. The admittances are

$$Y_R = \frac{1}{R}, \quad Y_C(s) = sC, \quad Y_L(s) = \frac{1}{sL}$$

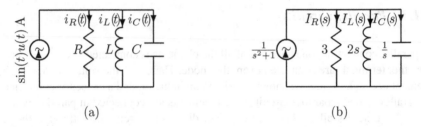

Fig. 3.4 Parallel circuit with a current source

Let $Y(s) = Y_R(s) + Y_L(s) + Y_C(s)$. Then, the currents are

$$I_R(s) = I(s)\frac{Y_R}{Y(s)} \quad I_L = I(s)\frac{Y_L(s)}{Y(s)} \quad I_C(s) = I(s)\frac{Y_C(s)}{Y(s)}$$

Consider the same circuit in frequency domain, shown in Fig. 3.4b with specific values for the components. The circuit is energized by a current source

$$i(t) = \sin(t)u(t) \leftrightarrow I(s) = \frac{1}{s^2 + 1}$$

The admittances are, with $R = 3\Omega$, $L = 2H$, and $C = 1F$,

$$Y_R = \frac{1}{3}, \quad Y_C = s \quad Y_L = \frac{1}{2s}$$

Let

$$Y(s) = Y_R(s) + Y_L(s) + Y_C(s) = \frac{1}{3} + s + \frac{1}{2s} = \frac{6s^2 + 2s + 3}{6s}$$

The currents are

$$I_R(s) = I(s)\frac{Y_R}{Y(s)} = \frac{2s}{(6s^2 + 2s + 3)(s^2 + 1)} \quad \text{and in the time domain}$$

$$i_R(t) = (0.5547 \cos(t - 2.5536) + 0.5708 e^{-0.1667t} \cos(0.6872t + 0.6290))u(t)$$

$$I_L(s) = I(s)\frac{Y_L}{Y(s)} = \frac{3}{(6s^2 + 2s + 3)(s^2 + 1)} \quad \text{and in the time domain}$$

$$i_L(t) = (0.8321 \cos(t + 2.1588) + 1.2108 e^{-0.1667t} \cos(0.6872t - 1.1797))u(t)$$

$$I_C(s) = I(s)\frac{Y_C}{Y(s)} = \frac{6s^2}{(6s^2 + 2s + 3)(s^2 + 1)} \quad \text{and in the time domain}$$

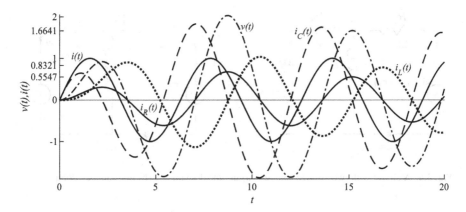

Fig. 3.5 The voltages and currents in the parallel circuit

$$i_C(t) = (1.6641\cos(t - 0.9828) + 1.2108e^{-0.1667t}\cos(0.6872t + 2.4378))u(t)$$

The voltages and currents in the parallel circuit are shown in Fig. 3.5. The input current is $i(t) = \sin(t)u(t)$. The currents $i_R(t)$, $i_L(t)$, and $i_C(t)$ are also shown. The voltage $v(t)$ across all the elements is the same, which is three times the current $i_R(t)$.

The phase of a sinusoidal waveform of a certain frequency is determined by the occurrence of its nearest positive peak to $t = 0$. For example, the phase of the cosine wave is zero radians. The peaks of the inductor current $i_L(t)$ occur after those of the voltage $v(t)$ across it, as can be seen from the figure. The current lags the voltage. When a voltage is applied to the inductor, it opposes the change in current. The current builds up more slowly than the voltage resulting in a lag. The peaks of the capacitor current $i_C(t)$ occur ahead of those of the voltage $v(t)$ across it, as can be seen from the figure. The current leads the voltage. When a voltage is applied to the capacitor, current has to flow to charge it to raise the voltage across it. The voltage builds up more slowly than the current resulting in a lead. The phase difference is zero in resistive circuits, called in phase. The peaks of $i_R(t)$ and $v(t)$ occur at the same time. Circuits with lag and lead characteristics are important in the design of compensators in the design of control systems.

Impedances Connected in Series and Parallel

The analysis of series and parallel circuits is relatively straightforward. In general, most circuits are a combination of series and parallel circuits or connected in a random configuration in which none of the elements are in series or parallel. Obviously, combinations of the concepts of series and parallel circuits are used to analyze series–parallel circuits. Analysis of circuits with random configurations is presented next.

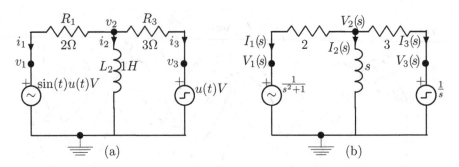

Fig. 3.6 A circuit with two voltage sources: (**a**) in time domain and (**b**) in frequency domain

3.1.4 Examples of Circuit Analysis

We analyze arbitrarily connected circuits, in circuit theory, with loop and nodal analysis. In these methods, the loop currents and node-to-datum voltages are selected as the independent variables.

Example 3.1 Consider the circuit shown in Fig. 3.6 with two voltage sources, two resistors, and an inductor. Assume that the initial current through the inductor is zero. Analyze the circuit.

Solution

$$R_1 = 2\Omega, \quad L_2 = 1H, \quad R_3 = 3\Omega, \quad v_1 = \sin(t)u(t)\,V, \quad v_3 = u(t)V$$

In the Laplace or frequency domain, we get

$$R_1 = 2, \quad Z_2 = s, \quad R_3 = 3, \quad V_1(s) = \frac{1}{s^2+1} \quad V_3(s) = \frac{1}{s}$$

Voltage $V_2(s)$ is the only unknown. Therefore, applying Kirchhoff's current law at node 2, we get

$$\frac{(V_2(s) - \frac{1}{s^2+1})}{R_1} + \frac{\left(V_2(s) - \frac{1}{s}\right)}{R_3} + \frac{V_2(s)}{Z_2} = 3s\left(V_2(s) - \frac{1}{s^2+1}\right)$$

$$+2s\left(V_2(s) - \frac{1}{s}\right) + 6V_2(s) = 0$$

Simplifying, we get

$$V_2(s)(5s+6) = \frac{3s}{s^2+1} + 2$$

Solving for $V_2(s)$, we get

$$V_2(s) = \frac{3}{5}\frac{s}{(s^2+1)(s+\frac{6}{5})} + \frac{2}{5}\frac{1}{(s+\frac{6}{5})} = \frac{3}{5}\frac{s}{(s+j)(s-j)(s+\frac{6}{5})} + \frac{2}{5}\frac{1}{(s+\frac{6}{5})}$$

$$= \frac{3}{5}\frac{0.2459-j0.2049}{(s-j)} + \frac{3}{5}\frac{0.2459+j0.2049}{(s+j)} + \frac{2}{5}\frac{1}{(s+\frac{6}{5})} + \frac{3}{5}\frac{(-0.4918)}{(s+\frac{6}{5})}$$

$$= \frac{0.1475-j0.1229}{(s-j)} + \frac{0.1475+j0.1229}{(s+j)} + \frac{0.1049}{(s+\frac{6}{5})}$$

Taking the inverse Laplace transform, we get

$$v_2(t) = (0.3840\cos(t-0.6947) + 0.1049e^{-1.2t})u(t)$$

The currents are

$$i_1(t) = 0.5(0.3840\cos(t-0.6947) - \sin(t) + 0.1049e^{-1.2t})u(t)$$

$$i_2(t) = \int_0^t v_2(\tau)d\tau = (0.3841\sin(t-0.6947) - 0.0874e^{-1.2t} + 0.3333)u(t)$$

$$i_3(t) = \frac{1}{3}(0.3840\cos(t-0.6947) - u(t) + 0.1049e^{-1.2t})u(t)$$

With $I_1 + I_2 + I_3 = 0$, Kirchhoff's current law is satisfied verifying the solution. This method of circuit analysis uses the internal description of the system.

The use of the linearity property in analyzing complex circuits makes the analysis easier. Using the linearity property, we can analyze the circuit with each excitation separately, find the responses, and add them to find the total response. In this procedure, when the response to one source is considered, the rest of the voltage sources are replaced by short-circuits and current sources are replaced by open-circuits. The internal resistances, if any, of the sources remain connected to the circuit. First, let us find the response to the unit-step signal alone.

$$3s(V_2(s)) + 2s\left(V_2(s) - \frac{1}{s}\right) + 6V_2(s) = 0$$

$$V_2(s) = \frac{0.4}{(s+\frac{6}{5})}$$

Taking the inverse Laplace transform, we get

$$v_2(t) = (0.4e^{-1.2t})u(t)$$

$$i_1(t) = (0.2e^{-1.2t})u(t)$$

$$i_2(t) = \int_0^t v_2(\tau)d\tau = \frac{1}{3}(1 - e^{-1.2t})u(t)$$

$$i_3(t) = \frac{1}{3}(0.4e^{-1.2t} - 1)u(t)$$

Now, let us find the response to the input $\sin(t)u(t)$ alone.

$$3s\left(V_2(s) - \frac{1}{s^2 + 1}\right) + 2s(V_2(s)) + 6V_2(s) = 0$$

$$V_2(s)(5s + 6) = \frac{3s}{s^2 + 1}$$

$$V_2(s) = \frac{3}{5}\frac{s}{(s^2 + 1)(s + \frac{6}{5})} = \frac{3}{5}\frac{s}{(s + j)(s - j)(s + \frac{6}{5})}$$

$$= \frac{3}{5}\frac{0.2459 - j0.2049}{(s - j)} + \frac{3}{5}\frac{0.2459 + j0.2049}{(s + j)} - \frac{(0.2951)}{(s + \frac{6}{5})}$$

$$= \frac{0.1475 - j0.1229}{(s - j)} + \frac{0.1475 + j0.1229}{(s + j)} - \frac{(0.2951)}{(s + \frac{6}{5})}$$

Taking the inverse Laplace transform, we get

$$v_2(t) = (0.3840\cos(t - 0.6947) - 0.2951e^{-1.2t})u(t)$$

$$i_1 = 0.5(0.3840\cos(t - 0.6947) - \sin(t) - 0.2951e^{-1.2t})u(t)$$

$$i_2(t) = \int_0^t v_2(\tau)d\tau = (0.3841\sin(t - 0.6947) + 0.2459e^{-1.2t})u(t)$$

$$i_3 = \frac{1}{3}(0.3840\cos(t - 0.6947) - 0.2951e^{-1.2t})u(t)$$

$$I_3 = -(I_1 + I_2)$$

The various responses of the circuit are shown in Fig. 3.7. The total responses $v_2(t)$, $i_1(t)$, $i_2(t)$, and $i_3(t)$ are shown, respectively, in (a), (d), (g), and (j). In the rest of the figures, the components of the responses obtained using the linearity property are shown. The sum of the components adds up to the total response.

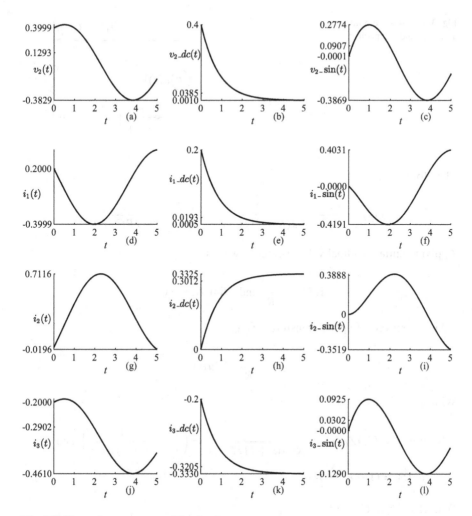

Fig. 3.7 The various responses of the circuit

RC Circuit

Consider the resistor–capacitor circuit shown in Fig. 3.8. It is a series circuit with a 2Ω resistor, a $0.5F$ capacitor, and the unit-step voltage source, $u(t)$. Let the current through the circuit be $i(t)$ and the voltages across the resistor and capacitor be, respectively, $v_R(t)$ and $v_C(t)$. Let the initial voltage across the capacitor be zero. Let us analyze the circuit.

The excitation and the impedance, in the Laplace transform domain, are

$$V(s) = \frac{1}{s} \quad \text{and} \quad Z(s) = R + (1/Cs)$$

Fig. 3.8 A series circuit with
a resistor and a capacitor

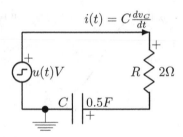

Therefore,

$$I(s) = \frac{V(s)}{Z(s)} = \frac{1}{s(R + (1/Cs))} = \frac{1}{R}\left(\frac{1}{s + (1/RC)}\right)$$

Applying initial and final value theorems, we get

$$i(0^+) = \frac{1}{R} \quad \text{and} \quad \lim_{t \to \infty} i(t) = 0$$

Taking the inverse Laplace transform, we get

$$i(t) = \frac{1}{R}e^{-\frac{t}{RC}}u(t)$$

With $Z_C = 1/Cs$,

$$V_C(s) = I(s)Z_C(s) = \frac{1}{RCs(s + (1/RC))} = \left(\frac{1}{s} - \frac{1}{(s + 1/RC)}\right) \quad \text{and}$$

$$v_C(t) = (1 - e^{-\frac{t}{RC}})u(t)$$

With $R = 2$ and $C = 0.5$,

$$i(t) = 0.5e^{-t}u(t)$$

$$v_C(t) = (1 - e^{-t})u(t)$$

Initial Condition
Let the initial voltage across the capacitor be $v_C(0) = 3V$. Find the current $i_0(t)$
with the input to the circuit is zero. That is the zero-input response of the circuit.

With magnitude $3\,V$ representing the initial condition, the excitation voltage due
to the initial condition and the impedance of the circuit, in the Laplace transform
domain, are

$$-\frac{v_C^-(0)}{s} \quad \text{and} \quad R + \frac{1}{Cs}$$

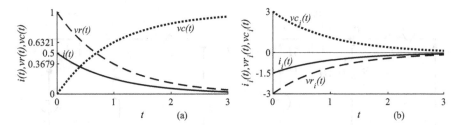

Fig. 3.9 The response of the series RC circuit: (**a**) to the unit-step voltage input alone and (**b**) to the initial voltage $3V$ across the capacitor alone

Dividing the voltage by the impedance, we get

$$I(s) = -\frac{v_C^-(0)}{s(R + \frac{1}{Cs})} = -\frac{v_C^-(0)}{R}\left(\frac{1}{s + \frac{1}{RC}}\right)$$

Applying the initial and final value theorems, we get the initial and final values of current, respectively, as

$$-\frac{v_C^-(0)}{R} \qquad \text{and} \qquad 0$$

Taking the inverse Laplace transform, we get

$$i(t) = -\frac{v_C^-(0)}{R}e^{-\frac{t}{RC}}u(t)$$

The initial and final currents are $i_0^+ = -v_C^-(0)/R$ and $i_\infty = 0$.

$$v_C(t) = v_C^-(0)(1 - e^{-\frac{t}{RC}})u(t)$$

Multiplying current $i(t)$ by the resistance R, we get

$$v_R(t) = -v_C^-(0)(e^{-\frac{t}{RC}})u(t)$$

The responses of the series RC circuit, with $R = 2\Omega$ and $C = 0.5F$, to the input alone and to the initial capacitor voltage 3 V alone are shown in Fig. 3.9a and b, respectively. The response of the RC circuit with input and initial voltages can be found using the linearity property.

Example 3.2 Consider the circuit shown in Fig. 3.10 with a unit-step voltage source, a resistor, a capacitor, and an inductor. Assume zero initial conditions. Analyze the circuit.

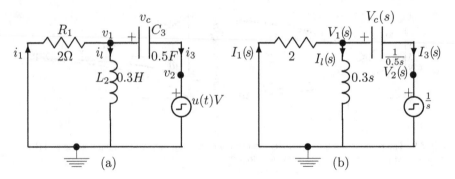

Fig. 3.10 A circuit: (**a**) in time domain and (**b**) in frequency domain

$$R_1 = 2\Omega, \quad L_2 = 0.3H, \quad C_3 = 0.5F, \quad v_2 = u(t)V$$

In the Laplace or frequency domain, we get

$$R_1 = 2, \quad Z_2 = 0.3s, \quad Z_3 = \frac{2}{s}, \quad V_2(s) = \frac{1}{s}$$

Voltage $V_1(s)$ is the only unknown. Therefore, applying Kirchhoff's current law at node 1, we get

$$\frac{V_1(s)}{R_1} + \frac{(V_1(s) - \frac{1}{s})}{Z_3} + \frac{V_1(s)}{Z_2} = \frac{V_1(s)}{2} + s\frac{(V_1(s) - \frac{1}{s})}{2} + \frac{V_1(s)}{0.3s} = 0$$

$$V_1(s) = \frac{0.3s}{0.3s^2 + 0.3s + 2}$$

Taking the inverse transform, we get

$$v_1(t) = (1.0193e^{-0.5t}\cos(2.5331t + 0.1949))u(t)$$

$$i_1(t) = -0.5v_1(t)$$

$$i_l(t) = \frac{1}{(0.3s^2 + 0.3s + 2)} = (1.3159e^{-0.5t}\cos(2.5331t - 0.5\pi))u(t)$$

$$i_C(t) = i_1(t) - i_l(t)$$

$$-i_1(t) + i_l(t) + i_3(t) = 0$$

3.2 Summary

- An electrical system is an interconnection of components such as resistors, inductors, and capacitors to produce, transfer, and utilize electrical power.
- For the analysis of any type of systems, a mathematical model has to be derived with some assumptions, so that the complexity of the model depends on the accuracy required.
- Using the voltage–current relationships of various components and the basic laws of interconnection, the desired outputs are found when the inputs are specified.
- The current flowing through a resistor is directly proportional to the voltage applied across it and it is inversely proportional to its value at all instants of time.
- The voltage across a capacitor is the time integral of the current flowing through it times the reciprocal of its value.
- The voltage across an inductor is the time derivative of the current flowing through it times its value.
- In a series connection, one, and only one, terminal of an element is connected to adjoining elements.
- In parallel connection, one terminal of all the elements is connected to one node and the other terminals are connected to another node.
- Kirchhoff's voltage law (KVL): The algebraic sum of voltages across all branches around any closed loop of a circuit is zero at all instants of time.
- Kirchhoff's current law (KCL): The algebraic sum of all currents at any node leaving through branches is zero at all instants of time.

Exercises

3.1 Determine the current in the RL circuit, shown in Fig. 3.11. The initial current through the inductor is $i(0^-) = 1A$ and the input is $x(t) = 2u(t)\ V$, the step signal. Determine also the current in the RL circuit with zero initial condition.

*3.2 Determine the current in the RC circuit, shown in Fig. 3.12. The initial voltage across capacitor is $v(0^-) = 1V$ and the input $x(t) = 2u(t)V$, the step signal. Determine also the current in the RC circuit with zero initial conditions.

Fig. 3.11 An RL circuit in time domain.

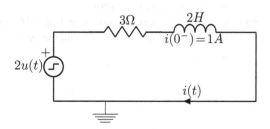

Fig. 3.12 An *RC* circuit in time domain.

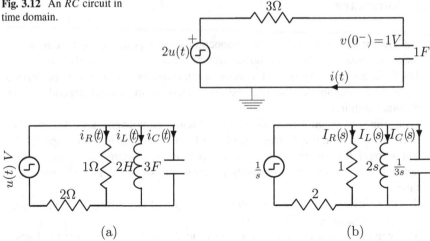

Fig. 3.13 Series–parallel circuit with a unit-step voltage source: (**a**) time-domain representation and (**b**) frequency-domain representation

Fig. 3.14 An *RLC* circuit in time domain

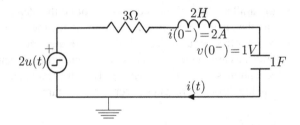

3.3 Determine the current in the series–parallel circuit, shown in Fig. 3.13. The input, $x(t) = u(t)$ V, is a voltage source. Assume zero initial conditions.

*3.4 Determine the current in the *RLC* circuit, shown in Fig. 3.14. The initial current through the inductor is $i(0^-) = 2A$ and the initial voltage across capacitor $v(0^-) = 1V$ and the input $x(t) = 2u(t)V$, the step signal. Determine also the current in the *RLC* circuit with zero initial conditions.

3.5 Consider the circuit shown in Fig. 3.15 with two voltage sources, two resistors, and a capacitor. Assume that the initial voltage across the capacitor is zero. Analyze the circuit.

*3.6 Consider the circuit shown in Fig. 3.16 with a unit-step voltage source, a resistor, a capacitor, and an inductor. Assume zero initial conditions. Analyze the circuit.

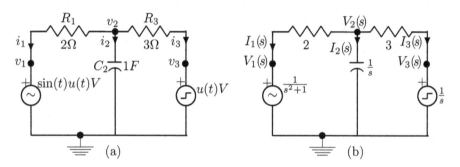

Fig. 3.15 A circuit with two voltage sources: (**a**) in time domain and (**b**) in frequency domain

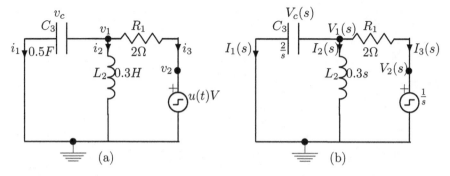

Fig. 3.16 A circuit: (**a**) in time domain and (**b**) in frequency domain

Chapter 4
Mathematical Modeling of Mechanical Systems

In order to analyze and design systems, the first step is to represent the system by a mathematical model. The mathematical model of a system is, usually, a set of differential equations that characterizes the physical system behavior with acceptable accuracy. For example, the basic components of mechanical systems are mass, spring, and damper, shown on the right side of Fig. 4.1. They are assumed to be ideal and linear in the operating range. With their characteristic equations and Newton's laws, we can model any linear mechanical system adequately.

Physically different systems, such as mechanical and electrical systems, have similarity between their equilibrium equations. They are called analogous systems. This implies that electrical systems, which are easier to analyze, can be constructed whose behavioral characteristics are similar to those of the mechanical systems. As we are used to the vast electrical circuit theory, we can solve mechanical system analysis problem by the more convenient electrical circuit analysis. All engineering systems can be adequately modeled using a few basic elements with their input–output characteristics known and some laws governing their interconnection. For example, electrical systems can be modeled using ideal resistors, inductors, and capacitors, which are shown on the left side of Fig. 4.1. Even with the idealization, it is almost always possible to find a model of a system that adequately represents physical systems with acceptable tolerances. In this chapter, we present modeling of mechanical systems. Other systems, such as hydraulic and thermal systems, can also be represented by electrical analogs.

4.1 Modeling Electrical Systems

Resistor is a dissipative element, while the other two are storage elements. In modeling systems, we use through and across variables. A **through variable** is a variable that does not change between the ends of a circuit element, for example, the current flowing through a resistor. An **across variable** is a variable that changes

D. Sundararajan, *Control Systems*, https://doi.org/10.1007/978-3-030-98445-8_4

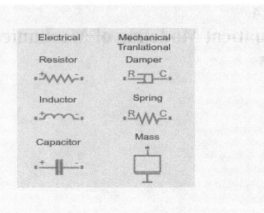

Fig. 4.1 Representation of basic electrical (on the left) and mechanical elements

between the ends of a circuit element, for example, the voltage at the ends of a resistor. A through variable, such as current, is measured by connecting the sensor in series with an element, such as resistor. An across variable, such as voltage, is measured by connecting the sensor in parallel with an element, such as resistor. In electrical circuits, the through variable is the current $i(t)$ and the voltage $v(t)$ or $e(t)$ is the across variable. The resistor, inductor, and capacitor are characterized by the equations

$$v = iR, \qquad v = L\frac{di}{dt}, \qquad i = C\frac{dv}{dt} \tag{4.1}$$

where v and i are arbitrary functions of time except that they are related to each other by Eq. (4.1). In modeling other type of systems, the values of R, L, and C are specified in terms of the parameters of the other systems. Commonly used symbols and units in electrical engineering are shown in Table 4.1.

4.2   Modeling Translational Mechanical Systems

The topology of a network is the way the various components of a system are interconnected. There are two types of analogies those are commonly used. Here, we present the one that preserves the topologies of the network of different systems those can be transformed with a one-to-one correspondence that is continuous in both directions. As voltage and velocity are across variables and current and force are through variables, the topology of the systems remains the same using this analogy. Learning both types of analogies at the same time is difficult. After getting used to one type, the other, if required, can be learned easily.

Table 4.1 Symbols and units in electrical engineering. Lowercase symbols imply a time variable quantity and uppercase symbols indicate a constant

Quantity	Symbol	Unit	Unit abbreviation
Charge	q	Coulomb	C
Current	i, I	Ampere	A
Flux linkages	ψ	Weber-turn	Wb
Energy	w, W	Joule	J
Voltage	v, V	Volt	V
Power	p, P	Watt	W
Capacitance	C	Farad	F
Inductance	L, M	Henry	H
Resistance	R	Ohm	Ω
Time	t	Second	sec
Cyclic Frequency	f	Hertz	Hz
Angular Frequency	ω	Radian/second	rad/sec

The relations between displacement x (magnitude and direction of motion), velocity v (rate of change of displacement with time in a given direction), and acceleration a (rate of change of velocity) are

$$v = \frac{dx}{dt} \quad \text{and} \quad a = \frac{dv}{dt} = \frac{d^2x}{dt}$$

Translational mechanical systems (which move back and forth in a straight line) can be modeled using ideal masses, springs, and dampers, which are shown on the right side of Fig. 4.1. Damper is a dissipative element that dissipates energy. The linear spring stretches or compresses proportional to the applied force. The spring stores potential energy. The mass M of a body of weight W is defined as

$$M = \frac{W}{g}$$

where g is the acceleration of free fall of the body due to gravity $g = 9.8066 \text{m/s}^2$. The mass stores kinetic energy.

In mechanical systems, the through variable is the force $f(t)$ and the velocity $v(t)$ is the across variable. The damper of viscous friction type, linear spring, and mass are characterized by the equations

$$f = Bv, \qquad f = Kx, \qquad f = M\frac{dv}{dt} = Ma \tag{4.2}$$

where v and f are arbitrary functions of time except that they are related to each other by Eq. (4.2). In using electrical analog of translational mechanical systems, the values of R, L, and C are specified as $R = 1/B$, $L = 1/K$, and $C = M$. Commonly used symbols and units in mechanical engineering are shown in Table 4.2. Analogy

Table 4.2 Symbols and units in mechanical engineering

Quantity	Symbol	Unit	Unit Abbreviation
Mass	M	Kilogram	kg
Displacement	x	Meter	m
Velocity	v	Meter/second	m/sec
Acceleration	a	Meter/second^2	m/sec^2
Angular Displacement	θ	Radian	rad
Angular Velocity	ω	Radian/second	rad/sec
Angular Acceleration	$\alpha = \frac{d\omega}{dt}$	Radian/second^2	rad/sec^2
Spring	K	Newton/meter	N/m
Friction	B	Newton/(meter/second)	N/(m/sec)
Force	F	Newton	N
Torque	τ	Newton-meter	N-m
Inertia	J	Kilogram-meter^2	Kg-m^2

Table 4.3 Analogy of translational mechanical system and electrical system

Analogy Description	Translational mechanical system	Electrical system
Through variable	Force, f	Current, i
Across variable	Velocity, v	Voltage, v
Dissipative element	Frictional coefficient, B	Resistance, $R = \frac{1}{B}$
	$v = \frac{f}{B}, \quad \frac{f^2}{B} = \frac{v^2}{1/B}$	$v = iR, \quad \frac{v^2}{R} = i^2 R$
Storage element	Spring constant, K	Inductance, $L = \frac{1}{K}$
	$v = \frac{1}{K}\frac{df}{dt}, \quad E = 0.5\frac{f^2}{K}$	$v = L\frac{di}{dt}, \quad E = 0.5Li^2$
Storage element	Mass, M	Capacitance, $C = M$
	$f = M\frac{dv}{dt}, \quad E = 0.5Mv^2$	$i = C\frac{dv}{dt}, \quad E = 0.5Cv^2$
	Displacement, x	Magnetic flux, ψ

of translational mechanical system and electrical system is shown in Table 4.3. By Ohm's law, $v = iR$. Applying the analogy, for the mechanical system, we get $v = f/B$ by comparison. Similarly, the energy dissipated by the resistor is $i^2 R$ and that by a damper is f^2/B. The time integral of voltage is the magnetic flux linkage. The time integral of current is the electric charge. The instantaneous energy stored in the magnetic field of an inductance of L H is $0.5Li^2$, where $i(t)$ is the current flowing through the inductance. The instantaneous energy stored in the electric field of a capacitor of C F is $0.5Cv^2$, where $v(t)$ is the voltage across the capacitor. Similarly, the instantaneous energy stored by a spring is $0.5f^2/K$. The instantaneous energy stored by a mass is $0.5Mv^2$.

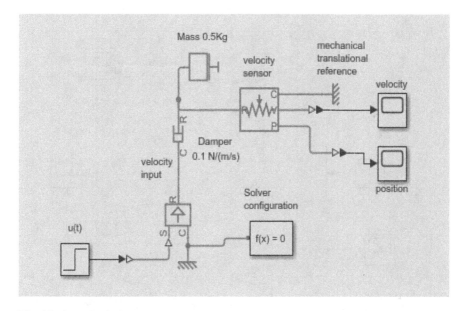

Fig. 4.2 A mechanical system

Model 1

The theory of control systems is highly mathematical. The easiest way to understand it is to use analytical methods along with simulation of the physical systems as given and also its electrical analog. The purpose of the electrical analog is to use the vast and well-developed electrical circuit theory, which we have studied in earlier courses.

Consider a relaxed mechanical system consisting of a damper and a mass, shown in Fig. 4.2. The system is shown as simulated in MATLAB® environment. The approach is to simulate physical systems using physical networks. The input is a velocity source. These simulations mimic physical systems. The readers are free to use other software packages of their choice. The input is a unit-step function, $u(t)$. This signal is converted to velocity input by the ideal translational velocity source block. This input is applied to the damper and the mass (with no friction) connected in series. The velocity across the mass is sensed by the ideal translational motion sensor. Ports V and P are, respectively, signals corresponding to velocity and position, which are displayed by the corresponding scopes connected. The solver configuration is a requirement in MATLAB software package. The C inputs of both the sensor and source blocks are connected to mechanical ("ground") translational reference. Irrespective of the force applied to it, the velocity remains zero (immovable position and velocity $= 0$). The problem is to find the velocity and the position of the mass. Let us say the input to the system is applied at time $t = 0$. That is, a constant velocity of 1 m/s is applied for $t \geq 0$. Due to inertia of the mass (the tendency of a body to maintain its state of rest), the velocity of the mass starts at 0 and increases exponentially to 1, with a time constant M/B.

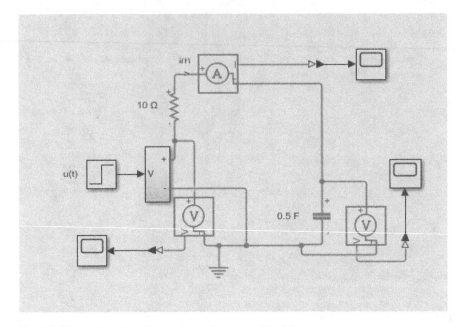

Fig. 4.3 Electrical analog of the mechanical system in Fig. 4.2

The task of finding the values of the variables of interest can be easily done using the more familiar corresponding electrical analog. The analog is obtained by replacing elements in Fig. 4.2 by their electrical analogs. The general procedure is to replace:

- Force generators by current sources
- Input velocities by voltage sources
- Friction elements by resistors
- Springs by inductors
- Masses by capacitors (which must be grounded). The energy of a capacitor must be measured relative to electrical ground
- Fixed reference by ground

If capacitors are not grounded, the analogy becomes difficult to apply. The voltage at an electrical ground remains zero, irrespective of the current flow. The electrical analog of the mechanical system in Fig. 4.2 is shown in Fig. 4.3.

4.2.1   Theoretical Analysis

The differential equation characterizing the circuit shown in Fig. 4.3, with unit-step voltage input $u(t)$, is

$$\frac{1}{C} \int_0^t i(t) + Ri(t) = u(t)$$

Differentiating this equation with respect to t and dividing both sides by R, with zero initial condition, we get

$$\frac{di(t)}{dt} + \frac{1}{RC}i(t) = \frac{1}{R}\delta(t) \tag{4.3}$$

The derivative of the unit-step function is the unit-impulse function. While we can solve the differential equation in the time domain, using the Laplace transform is much easier as the differential equation reduces to an algebraic equation. Taking the Laplace transform, we get

$$I(s)\left(s + \frac{1}{RC}\right) = \frac{1}{R}$$

Solving for $I(s)$, we get

$$I(s) = I_C(s) = \frac{1}{R}\left(\frac{1}{s + (1/RC)}\right)$$

The value of the resistor is $R = \frac{1}{0.1} = 10\Omega$ and that of the capacitor is $C = 0.5F$, as shown in Fig. 4.3. The initial voltage across the capacitor be zero. Taking the inverse Laplace transform, we get

$$i(t) = i_C(t) = \frac{1}{R}e^{-\frac{t}{RC}}u(t)$$

Applying initial and final value theorems to the expression for $I(s)$, we get

$$i(0^+) = \frac{1}{R} \quad \text{and} \quad \lim_{t \to \infty} i(t) = 0$$

which is in accordance with the complete solution. With $Z_C = 1/Cs$,

$$V_C(s) = I(s)Z_C(s) = \frac{1}{RCs(s + (1/RC))}$$

$$= \left(\frac{1}{s} - \frac{1}{(s + 1/RC)}\right) \quad \text{and} \quad v_C(t) = (1 - e^{-\frac{t}{RC}})u(t)$$

The unit-step response is

$$v_C(t) = (1 - e^{-\frac{t}{RC}})u(t) \quad \text{and} \quad i_C(t) = \frac{1}{R}e^{-\frac{t}{RC}}u(t)$$

When the step input is applied, the initial voltage across the capacitor is zero. Therefore, the current through the circuit is limited only by the resistor. As the

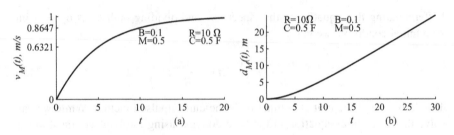

Fig. 4.4 (a) The velocity of the mass and (b) the distance the mass has moved

capacitor charges to the input voltage, the input current exponentially decays to zero.

As voltage corresponds to velocity in the mechanical system, we have to integrate the voltage to find the distance the mass moved in the mechanical system. By integrating $V_C(s)$, we get

$$\frac{V_C(s)}{s} = \left(\frac{1}{s^2} - \frac{1}{s(s+1/RC)}\right) = \left(\frac{1}{s^2} - RC\left(\frac{1}{s} - \frac{1}{(s+1/RC)}\right)\right)$$

By taking the inverse Laplace transform, we get

$$\left(t - RC + RCe^{-\frac{t}{RC}}\right)u(t)$$

Now, we have found out the values of the variables in the mechanical system using its electrical analog. Figure 4.4a and b show, respectively, the velocity of the mass and the distance the mass has moved.

The derivative of these expressions, the unit-impulse response, is

$$v_C(t) = \frac{1}{RC}e^{-\frac{t}{RC}}u(t) \quad \text{and} \quad i_C(t) = -\frac{1}{R^2C}e^{-\frac{t}{RC}}u(t)$$

■

For the mechanical system, the equilibrium equation is

$$\frac{1}{M}\int_{-\infty}^{t} f(t) + \frac{1}{B}f(t) = u(t)$$

Differentiating with respect to t, we get

$$\frac{df}{dt} + \frac{B}{M}f(t) = B\delta(t) \tag{4.4}$$

By analogy, $f(t) \leftrightarrow i(t)$, $M \leftrightarrow C$, and $B \leftrightarrow \frac{1}{R}$. Therefore, by applying the analogy, Eq. (4.4) becomes the same as Eq. (4.3).

Taking the Laplace transform of Eq. (4.4), we get

$$sF(s) + \frac{B}{M}F(s) = B$$

Solving for $F(s)$, we get

$$F(s) = \frac{B}{s + \frac{B}{M}}$$

Taking the inverse Laplace transform, we get

$$f(t) = Be^{-\frac{B}{M}t}u(t)$$

The velocity of the mass is

$$v(t) = \frac{B}{M}\int_0^t e^{-\frac{B}{M}\tau}d\tau = (1 - e^{-\frac{B}{M}t})u(t)$$

Integrating $v(t)$, we get the distance $d(t)$ as

$$d(t) = \left(t - \frac{M}{B} + \frac{M}{B}e^{-\frac{B}{M}t}\right)u(t)$$

The mechanical and electrical systems shown in Figs. 4.2 and 4.3 can be run in MATLAB and verify that the responses are the same. Usually the mechanical system is given and the task is to find the unknown mechanical variables. Convert it to the corresponding electrical analog and apply the circuit theory to find the unknown electrical variables. Then, apply the analogy to find the values of the corresponding unknown mechanical variables. For electrical systems, the algebraic sum of currents at a node is zero. Similarly, for mechanical systems, the algebraic sum of forces at an object is zero.

In electrical systems, with N number of resistors connected in series, the equivalent resistance R_{eq} of the series circuit is

$$R_{eq} = R_1 + R_2 + \cdots + R_N$$

The value of R_{eq} will be larger than the largest resistor in the series connection. The same current I passes through all the resistors. Therefore, the voltage V across the series connection is

$$V = IR_1 + IR_2 + \cdots + IR_N = IR_{eq}$$

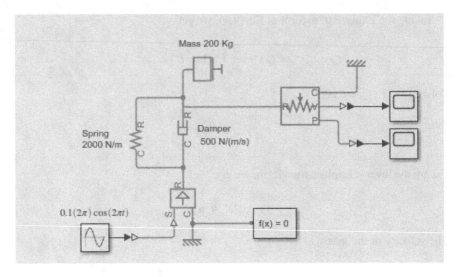

Fig. 4.5 A mechanical system

With N number of resistors connected in parallel, the equivalent resistance R_{eq} of the parallel circuit is

$$\frac{1}{R_{eq}} = \frac{1}{R_1} + \frac{1}{R_2} + \cdots + \frac{1}{R_N}$$

The value of R_{eq} will be smaller than the smallest resistor in the parallel connection, since the total current is more. The same voltage V is applied across all the resistors. As the topology does not change, the same rules apply for mechanical systems as well.

Model 2

Figure 4.5 shows a much simplified model of a car suspension system. When the car moves along the road, the vertical displacements at the tires are the input to the suspension system, which has to reduce the displacements of the body. The equilibrium position of the body is when the input is zero and the displacement of the body is measured with respect to this position. Let the input be

$$x_{in}(t) = 0.1 \sin(2\pi t) \text{ m/sec.}$$

Let the output be $x_o(t)$. For translational systems, the algebraic sum of forces on a rigid body in a given direction is equal to the product of the mass of the body and its acceleration in the same direction is Newton's second law of motion; that is,

$$\Sigma F = Ma$$

Applying Newton's second law to the system, we get

$$M\frac{dx_o^2}{dt} + B\left(\frac{dx_o}{dt} - \frac{dx_{in}}{dt}\right) + K(x_o(t) - x_{in}(t)) = 0$$

Shifting the input terms to the right side, we get

$$M\frac{dx_o^2}{dt} + B\frac{dx_o}{dt} + Kx_o(t) = B\frac{dx_{in}}{dt} + Kx_{in}(t)$$

In this book, we assume that systems are causal. That is, the system output depends on the past and present inputs only. With this assumption, the one-sided Laplace can be used in the transform domain for the design and analysis of systems. Assuming the initial conditions are zero and taking the Laplace transform, we get

$$s^2 M X_o(s) + Bs X_o(s) + K X_o(s) = Bs X_{in}(s) + K X_{in}(s)$$

The transfer function is

$$\frac{X_o(s)}{X_{in}(s)} = \frac{Bs + K}{Ms^2 + Bs + K} \tag{4.5}$$

For this system. the input–output pair is $X_{in}(s)$ and $X_o(s)$. Let us differentiate this pair to get the velocity input–output pair $sX_{in}(s) = V_{in}(s)$ and $sX_o(s) = V_o(s)$. Even for this pair, the transfer function remains the same for a linear system. In general, what happens to the input in the input–output pair must also happen to the other. If one is differentiated, the other one becomes differentiated. If one is integrated, the other one becomes integrated. If one function is differentiated twice, the other function is also differentiated twice. The conclusion is that, for any linear mechanical system, the transfer function remains the same for displacement, velocity, and acceleration. Using this property, for example, we can apply the velocity input to the system and determine the displacement output by integrating the corresponding velocity output.

Let the displacement input be

$$0.1\sin(2\pi t) \leftrightarrow \frac{(0.1)2\pi}{s^2 + (2\pi)^2}$$

The derivative of the displacement input is velocity

$$(0.1)2\pi \cos(2\pi t) \leftrightarrow \frac{(0.1)2\pi s}{s^2 + (2\pi)^2}$$

This velocity input is applied to the mechanical system in the simulation diagram. Similarly, voltage input is applied in the electrical analog circuit diagram, shown in Fig. 4.6. The output is

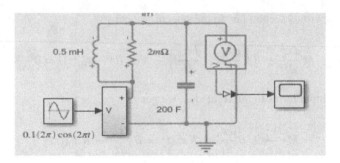

Fig. 4.6 Electrical analog of the mechanical system in Fig. 4.5

$$X_o(s) = \frac{0.1(2\pi)}{M} \frac{(Bs + K)}{\left(s^2 + \frac{B}{M}s + \frac{K}{M}\right)(s^2 + (2\pi)^2)}$$

System parameters are

$$M = 200\text{kg}, \qquad B = 500\text{N/(m/sec)}, \qquad K = 2000\text{N/m}$$

Substituting the numerical values, decomposing into partial fraction, and taking the inverse Laplace transform, we get

$$x_o(t) = (0.0648e^{-1.25t}\cos(2.9047t - 0.5389) + 0.0557\cos(2\pi t + 3.0643))u(t)$$

The first component of the output is the transient response, which becomes negligible after about four time constants of the system. What remains afterward is the steady-state component of the same form as the input sinusoid. Note that the magnitude of the input bumps due to road surface variations has been reduced to about one-half by the suspension system.

By analogy, $M \leftrightarrow C$, $B \leftrightarrow \frac{1}{R}$, and $K \leftrightarrow \frac{1}{L}$. Therefore, by applying the analogy, Eq. (4.5) is transformed into the transfer function of the corresponding electrical circuit,

$$\frac{X_o(s)}{X_{in}(s)} = \frac{\frac{1}{R}s + \frac{1}{L}}{Cs^2 + \frac{1}{R}s + \frac{1}{L}} = \frac{Ls + R}{RLCs^2 + Ls + R}$$

We can derive this result from the electrical circuit diagram, shown in Fig. 4.6, as well. The impedance of the parallelly connected resistor R and inductor L is

$$\frac{RLs}{R + Ls}$$

Therefore, the voltage across the capacitor C, by voltage division, is

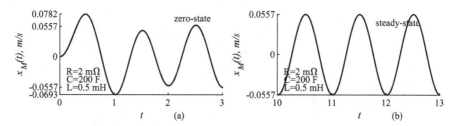

Fig. 4.7 (a) The zero-state response and (b) the steady-state response

$$V_o(s) = \frac{\frac{1}{sC}}{\frac{1}{sC} + \frac{RLs}{R+Ls}} V_{in}(s) = \frac{Ls + R}{RLCs^2 + Ls + R} V_{in}(s)$$

as derived earlier from the transfer function of the mechanical system using analogy. The electrical analog of the mechanical system in Fig. 4.5 is shown in Fig. 4.6. The displacement of the mass is shown in Fig. 4.7. In Fig. 4.7a, the zero-state response is shown. In Fig. 4.7b, the steady-state response is shown.

Model 3

Figure 4.8 shows a mechanical system. The input is force, $f(t) = \sin(4\pi t)$ Newton. For the mechanical system, the algebraic sum of forces at an object is zero, similar to KCL for electrical systems. Therefore,

$$F_{in}(s) = (sM + B)V(s) \quad \text{and} \quad \frac{V(s)}{F_{in}(s)} = \frac{1}{M} \frac{1}{\left(s + \frac{B}{M}\right)}$$

By analogy, $M \leftrightarrow C$, $B \leftrightarrow \frac{1}{R}$, and $F_{in}(s) \leftrightarrow I(s)$. Applying the analogy, we get the corresponding expression for the electrical analog of the system.

Figure 4.9 shows the corresponding electrical analog to the system in Fig. 4.8. The input current is $i(t) = \sin(4\pi t)$ A. The voltage across the capacitor is

$$V_C(s) = \frac{4\pi}{C} \frac{1}{(s^2 + (4\pi)^2)\left(s + \frac{1}{RC}\right)}$$

System parameters are

$$C = 0.1\text{F}, \qquad R = 2\Omega, \qquad L = 0.3\text{H}$$

Substituting the numerical values, decomposing into partial fraction, and taking the inverse Laplace transform, we get

$$v(t) = (0.6870e^{-5t} + 0.7394\cos(4\pi t - 2.7629))u(t)$$

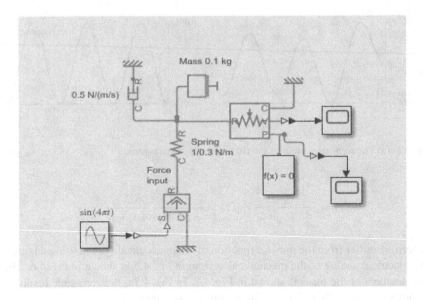

Fig. 4.8 A mechanical system

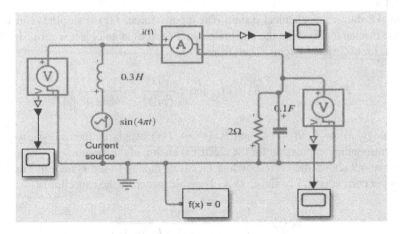

Fig. 4.9 The electrical analog of the mechanical system shown in Fig. 4.8

This voltage corresponds to velocity in the mechanical system. We have to integrate $v(t)$ to get the displacement of the mass $d(t)$. Integrating $v(t)$, we get

$$\frac{4\pi}{C} \frac{1}{s(s^2 + (4\pi)^2)\left(s + \frac{1}{RC}\right)}$$

Substituting the numerical values, decomposing into partial fraction, and taking the inverse Laplace transform, we get

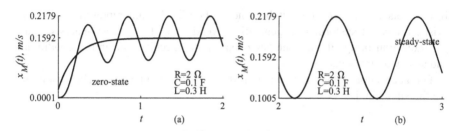

Fig. 4.10 (a) The zero-state response of the displacement of the mass and (b) the steady-state response

$$d(t) = (0.1592 - 0.1374e^{-5t} + 0.0588\cos(4\pi t + 1.9495))u(t)$$

The displacement of the mass is shown in Fig. 4.10. In Fig. 4.10a, the zero-state response is shown. In Fig. 4.10b, the steady-state response is shown.

Response to Initial Conditions
So far, we analyzed initially relaxed systems. That is, the initial conditions of the storage elements, such as capacitors and inductors, are assumed to be zero. In practice, the response of a system with nonzero conditions, when the input is applied, is also required. It turns out that the previous history of the circuit behavior can be summarized by a knowledge of the voltages across the capacitors and the currents through the inductors at the instsant the input is applied. Therefore, it is straightforward to extend the analysis for systems with nonzero initial values of its storage elements. The component with initial value can be represented by a relaxed component, in addition to an appropriate input source. Then, the response to the initial value and the given excitation can be computed and added due to the linearity property.

The input-output relationship of an inductor in the time domain, with initial current $i(0^-) = 0$, is

$$v(t) = L\frac{di(t)}{dt}$$

where $v(t)$ and $i(t)$ are, respectively, the voltage across and the current through the inductor of value L henries. In the Laplace domain, the independent variable for voltages and currents is the complex frequency s and the basis functions are of the form e^{st}. The derivative of this function with respect to t, se^{st}, is of the same form except that it is multiplied by the value of the frequency s. In taking the Laplace transform, the input waveform is decomposed in terms of frequency components of all frequency. Therefore, the representation of an inductor becomes an algebraic equation instead of a differential equation; that is,

$$V(s) = LsI(s)$$

If we replace s by $j\omega$, then it becomes a Fourier-domain representation, if it exists. Laplace transform is a generalized version of the Fourier analysis. This generalization brings the advantages of analyzing circuits with initial conditions and also with unbounded signals.

For a capacitor, with initial voltage $v(0^-) = 0$, the input–output relationship, in the time domain, is

$$v(t) = \frac{1}{C} \int_0^t i(t)\, dt$$

where $v(t)$ and $i(t)$ are, respectively, the voltage across and the current through the capacitor of value C farads. The representation of a capacitor, in the Laplace domain, becomes an algebraic equation instead of a differential equation; that is,

$$V(s) = \frac{1}{Cs} I(s)$$

For a resistor of value $R\,\Omega$, which has no storage capability,

$$V(s) = RI(s)$$

Summarizing, in the frequency domain, the voltage–current relationship of all the elements is algebraic, with their impedance values R, sL, and $1/Cs$. The impedance of an element is the ratio $V(s)/I(s)$, assuming zero initial conditions.

For a capacitor, with an initial voltage $v(0^-)$,

$$i(t) = C \left(\frac{dv(t)}{dt} - v(0^-)\delta(t) \right)$$

Taking the Laplace transform of this expression, we get

$$I(s) = C(sV(s) - v(0^-)) \quad \text{or} \quad V(s) = \frac{I(s)}{sC} + \frac{v(0^-)}{s}$$

The charged capacitor is represented as an impedance $\frac{1}{sC}$ in series with an ideal voltage source $\frac{v(0^-)}{s}$, which corresponds to a step voltage $v(0^-)u(t)$ in the time domain. The additional voltage source in series becomes a short circuit with no initial charge.

By rearranging the expression for $V(s)$, we get an alternative representation as

$$V(s) = \frac{1}{sC} \left(I(s) + Cv(0^-) \right)$$

The voltage across the capacitor is due to the current $(I(s) + Cv(0^-))$ flowing through it, impedance multiplied by the current. This form of representation implies an uncharged capacitor in parallel with an impulsive current source $Cv(0^-)$. There is an additional current source in parallel, which becomes an open circuit with no initial current.

In the time domain, an inductor is characterized by

$$v(t) = L\left(\frac{di(t)}{dt} - i(0^-)\delta(t)\right)$$

In the Laplace domain, we get

$$V(s) = L(sI(s) - i(0^-)) = LsI(s) - Li(0^-) = sL\left(I(s) - \frac{i(0^-)}{s}\right)$$

The inductor is modeled as an impedance sL in series with an ideal impulsive voltage source $-Li(0^-)$, which becomes a short circuit with no initial current.

Alternatively, the voltage across the inductor is due to the current $(I(s) - \frac{i(0^-)}{s})$ flowing through it. An inductor with initial current $i(0^-)$ is modeled as an inductor, with no initial current, in parallel with a current source $\frac{i(0^-)}{s}$, which becomes an open circuit with no initial current.

Model 4

Figure 4.11 shows a mechanical system with initial velocity of the mass 0.2 m/s and initial spring deformation 0.03 m with parameters

$$M = 0.1 \text{ Kg}, \qquad K = 1/0.3 \text{ m/sec}, \qquad B = 0.5 \text{ N/(m/s)}$$

The equivalent electrical analog is shown in Fig. 4.12. The initial capacitor voltage is set as 0.2 V and the initial current through the inductor i_0 is set as $0.03/0.3 = 0.1$ A. Note that $0.03 = Li_0 = 0.3i_0$. Now, the problem is to determine the current in the electrical analog RLC circuit, shown in Fig. 4.12. The initial current through the inductor is $i(0^-) = 0.1A$ and the initial voltage across the capacitor is $v(0^-) = 0.2V$. The input is $x(t) = u(t)V$, the unit-step signal. The values of R, C, and L are, respectively, 2Ω, 0.1 F, and 0.3 H.

Let us find the response with zero initial conditions. In the Laplace domain, the input voltage to the circuit is

$$\frac{1}{s}$$

The impedance of the circuit is

$$2 + 0.3s + \frac{10}{s} = \frac{0.3s^2 + 2s + 10}{s}$$

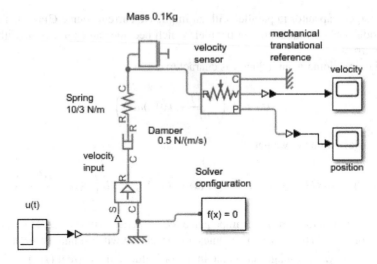

Fig. 4.11 A mechanical system

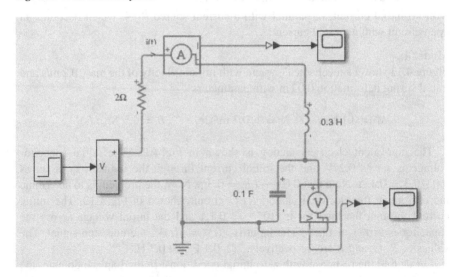

Fig. 4.12 The electrical analog of the mechanical system in Fig. 4.11

We get the current in the circuit by dividing the voltage by the impedance

$$I(s) = \frac{1}{0.3s^2 + 2s + 10} = \frac{(10/3)}{s^2 + (20/3)s + (100/3)}$$

Expanding into partial fractions, we get

$$I(s) = \frac{-j0.3536}{s + (3.3333 - j4.7140)} + \frac{j0.3536}{s + (3.3333 + j4.7140)}$$

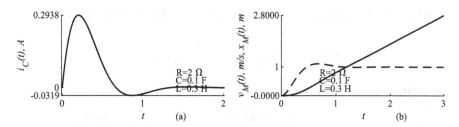

Fig. 4.13 (a) The current through the series RLC circuit with zero initial conditions and (b) the velocity (dashed line) and displacement of the mass

Expressing the numerators in polar form, we get

$$I(s) = \frac{0.3536\angle - \pi/2}{s + (3.3333 - j4.7140)} + \frac{0.3536\angle\pi/2}{s + (3.3333 + j4.7140)}$$

Taking the inverse Laplace transform, we get the current in the circuit as

$$i(t) = (0.7071e^{-3.3333t} \cos\left(4.7140t - \frac{\pi}{2}\right) u(t)$$

The current through the series RLC circuit with zero initial conditions is shown in Fig. 4.13a.

With the capacitive impedance $Z_C(s) = 1/sC$, the voltage across the capacitor $V_C(s)$ is

$$V_C(s) = I(s)Z_C(s) = \frac{(10/3)}{Cs(s^2 + (20/3)s + (100/3))}$$

Expanding into partial fractions, we get

$$V_C(s) = \frac{1}{s} + \frac{0.5 - j0.3536}{s + (3.3333 - j4.7140)} + \frac{0.5 + j0.3536}{s + (3.3333 + j4.7140)}$$

Expressing the numerators in polar form, we get

$$V_C(s) = \frac{1}{s} + \frac{0.6124\angle 2.5261}{s + (3.3333 - j4.7140)} + \frac{0.6124\angle - 2.5261}{s + (3.3333 + j4.7140)}$$

Taking the inverse Laplace transform, we get the voltage across the capacitor, which corresponds to velocity of the mass, as

$$v_C(t) = (1 + 1.2247e^{-3.3333t} \cos(4.7140t + 2.5261))u(t)$$

By integrating the velocity of the mass, we get the displacement of the mass as

$$D_M(s) = \frac{(10/3)}{Cs^2(s^2 + (20/3)s + (100/3))}$$

Expanding into partial fractions, we get

$$D_M(s) = \frac{1}{s^2} + \frac{1}{s} + \frac{0.1 + j0.0354}{s + (3.3333 - j4.7140)} + \frac{0.1 - j0.0354}{s + (3.3333 + j4.7140)}$$

Expressing the numerators in polar form, we get

$$D_M(s) = \frac{1}{s^2} - \frac{0.2}{s} + \frac{0.1061\angle 0.3398}{s + (3.3333 - j4.7140)} + \frac{0.6124\angle - 0.3398}{s + (3.3333 + j4.7140)}$$

Taking the inverse Laplace transform, we get the voltage across the capacitor, which corresponds to velocity of the mass, as

$$d_M(t) = (t - 0.2 + 0.2121e^{-3.3333t} \cos(4.7140t + 0.3398))u(t)$$

With nonzero initial conditions, the net voltage in the circuit is

$$\frac{1}{s} + 0.03 - \frac{0.2}{s} = \frac{0.8}{s} + 0.03$$

The impedance of the circuit is

$$2 + 0.3s + \frac{10}{s} = \frac{0.3s^2 + 2s + 10}{s}$$

We get the current in the circuit by dividing the voltage by the impedance

$$I(s) = \frac{0.03s + 0.8}{0.3s^2 + 2s + 10} = \frac{(0.1)s + (8/3)}{s^2 + \frac{20}{3}s + \frac{100}{3}}$$

Expanding into partial fractions, we get

$$I(s) = \frac{0.2525 - j1.3714}{s + (3.3333 - j4.7140)} + \frac{0.2525 + j1.3714}{s + (3.3333 + j4.7140)}$$

Expressing the numerators in polar form, we get

$$I(s) = \frac{0.505\angle - 1.3714}{s + (0.3333 - j0.4714)} + \frac{0.505\angle 1.3714}{s + (0.3333 + j0.4714)}$$

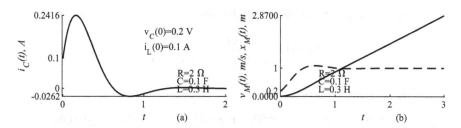

Fig. 4.14 (a) The current through the series RLC circuit with nonzero initial conditions and (b) the velocity (dashed line) and displacement of the mass

If two terms are conjugate in a partial-fraction expansion, then the inverse is twice the real part of any one of the inverses. For example,

$$I(s) = \frac{r \angle \theta}{s + (a - jb)} + \frac{r \angle -\theta}{s + (a + jb)}$$

Taking the inverse Laplace transform, we get the current in the circuit as

$$i(t) = 2re^{-at} \cos(bt + \theta)u(t)$$

For this example, we get

$$i(t) = 0.505e^{-3.3333t} \cos(4.7140t - 1.3714)u(t)$$

The current through the series RLC circuit with initial conditions is shown in Fig. 4.14.

The voltage across the capacitor is

$$V_C(s) = \frac{(0.1)s + (8/3)}{Cs \left(s^2 + \frac{20}{3}s + \frac{100}{3}\right)}$$

Expanding into partial fractions, we get

$$V_C(s) = \frac{0.8}{s} + \frac{-0.4000 + j0.1768}{s + (3.3333 - j4.7140)} + \frac{-0.4000 - j0.1768}{s + (3.3333 + j4.7140)}$$

Expressing the numerators in polar form, we get

$$V_C(s) = \frac{0.8}{s} + \frac{0.4373 \angle 2.7255}{s + (0.3333 - j0.4714)} + \frac{0.4373 \angle -2.7255}{s + (0.3333 + j0.4714)}$$

Taking the inverse Laplace transform, we get the voltage across the capacitor due to input, including the initial voltage 0.2, as

$$v_c(t) = (0.2 + 0.8 + 0.8746e^{-3.3333t} \cos(4.7140t + 2.7255))u(t)$$

Integrating the capacitor voltage, we get the displacement of the mass as

$$D_M(s) = \frac{(0.1)s + (8/3)}{Cs^2 \left(s^2 + \frac{20}{3}s + \frac{100}{3}\right)}$$

Expanding into partial fractions, we get

$$D_M(s) = \frac{0.8}{s^2} - \frac{0.13}{s} + \frac{0.0650 + j0.0389}{s + (3.3333 - j4.7140)} + \frac{0.0650 - j0.0389}{s + (3.3333 + j4.7140)}$$

Expressing the numerators in polar form, we get

$$D_M(s) = \frac{0.8}{s^2} - \frac{0.13}{s} + \frac{0.0757\angle 0.5392}{s + (0.3333 - j0.4714)} + \frac{0.0757\angle -0.5392}{s + (0.3333 + j0.4714)}$$

Taking the inverse Laplace transform, we get the displacement of the mass, including the initial value 0.2, as

$$d_M(t) = 0.2t + 0.8t - 0.13 + 0.1515e^{-3.3333t}\cos(4.7140t + 0.5392)$$

$$= (t - 0.13 + 0.1515e^{-3.3333t}\cos(4.7140t + 0.5392))u(t)$$

4.3 Modeling Rotational Mechanical Systems

The rotational motion of a body is its motion about a fixed axis. The torque (a twisting force, instead of force in translational systems) and angular velocity (instead of linear velocity) are the basic variables similar to current and voltage for electrical systems. Mass is replaced by rotational mass or moment of inertia, linear spring is replaced by torsional spring, and damper is replaced by torsional damper. The characterizing equations are similar to those of the translational systems. For rotational systems, the algebraic sum of torques about a fixed axis is equal to the product of the inertia and the angular acceleration about the axis is the version of Newton's second law of motion; that is,

$$\Sigma \tau = J\alpha$$

The analogy of rotational mechanical and electrical systems is shown in Table 4.4.

4.3.1 Simple Pendulum

Swing, which is often found in children's parks, is a mechanical device to support someone swinging back and forth. A circuit with an inductor and a capacitor connected in series also produces oscillations similar to the oscillation of the swing.

Table 4.4 Analogy of rotational mechanical system and electrical system

Analogy		
Description	Rotational mechanical system	Electrical system
Through variable	Torque, τ	Current, i
Across variable	Angular velocity, ω	Voltage, v
Dissipative element	Frictional coefficient, B_r	Resistance, $R = \frac{1}{B_r}$
	$\omega = \frac{\tau}{B_r}, \quad \frac{\tau^2}{B_r} = \frac{\omega^2}{1/B_r}$	$v = iR, \quad \frac{v^2}{R} = i^2 R$
Storage element	Spring constant, K_r	Inductance, $L = \frac{1}{K_r}$
	$\omega = \frac{1}{K_r}\frac{d\tau}{dt}, \quad E = 0.5\frac{\tau^2}{K_r}$	$v = L\frac{di}{dt}, \quad E = 0.5Li^2$
Storage element	Inertia, J	Capacitance, $C = J$
	$\tau = J\frac{d\omega}{dt}, \quad E = 0.5J\omega^2$	$i = C\frac{dv}{dt}, \quad E = 0.5Cv^2$
	Angular displacement, θ	Charge, q

Fig. 4.15 A simple pendulum

maximum potential energy
zero kinetic energy

maximum kinetic energy
zero potential energy

The motion of the swing about its mean position is called the simple harmonic motion. The characterizing equation of the motion of the simple pendulum, shown in Fig. 4.15, is nonlinear. In the introduction of the pendulum, assuming that the angle of the movement is small enough so that $\sin(\theta) \approx \theta$, the pendulum is modeled as a linear system.

An ideal simple pendulum is a point mass suspended by a massless string from some point about which it can swing back and forth. In a practical simple pendulum, the mass is a small metal sphere with a large mass. The length of the string is much longer than that of the radius of the sphere and its mass is relatively very small of that of the mass. When a pendulum is set in motion, it oscillates in a sinusoidal manner. The maximum distance the mass is displaced from its equilibrium position is the amplitude of oscillation. The time it takes to make one complete cycle of oscillation is its period T s and the cyclic, and angular frequencies are, respectively, $f = 1/T$ Hz and $\omega = 2\pi f$ rad/sec.

When the pendulum is in motion, there will be a restoring force that tends to move the pendulum back to its equilibrium position. As the pendulum overshoots the equilibrium position, there will be a restoring force with the direction reversed. The electrical analog of the simple pendulum is shown in Fig. 4.16.

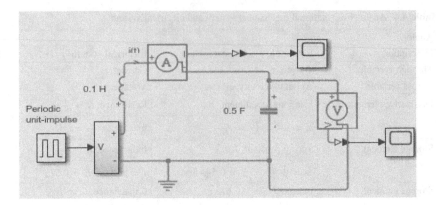

Fig. 4.16 The electrical analog of the simple pendulum in Fig. 4.15

The input voltage is an impulse in the time domain. In the Laplace domain, the input is 1. The circuit impedance is

$$Z(s) = sL + \frac{1}{sC} = \frac{1 + s^2 LC}{sC}$$

The current $I_C(s)$ in the circuit is

$$\frac{sC}{1 + s^2 LC} = \frac{s/L}{\frac{1}{LC} + s^2}$$

The voltage across the capacitor is

$$V_C(s) = \frac{sC}{1 + s^2 LC} \frac{1}{sC} = \frac{1}{1 + s^2 LC} = \frac{\frac{1}{LC}}{s^2 + \frac{1}{LC}}$$

The frequency of oscillation is $1/\sqrt{(LC)}$ rad/sec. Let the spring constant and inertia of the mechanical pendulum, respectively, be K_r and J. Then, with $L = 1/K_r = 0.1H$ and $C = J = 0.5F$,

$$V_C(s) = \frac{20}{s^2 + 20}$$

Taking the inverse transform, we get

$$v_C(t) = \sqrt{20} \sin(\sqrt{20}t) \quad \text{and} \quad i_C(t) = 10 \cos(\sqrt{20}t)$$

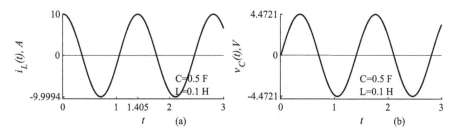

Fig. 4.17 (a) The current through the inductor and (b) the voltage across the capacitor

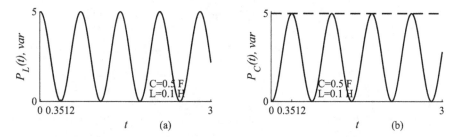

Fig. 4.18 (a) The instantaneous magnetic energy stored in the inductor and (b) the instantaneous electric energy stored in the capacitor

Figure 4.17a and b show, respectively, the current flowing through the inductor and the voltage across the capacitor. As they are, respectively, functions of cosine and sine, there is a 90° phase difference between them.

Figure 4.18a and b show, respectively, the instantaneous magnetic energy stored in the inductor and the instantaneous electric energy stored in the capacitor. In simple harmonic motion, the restoring force is directly proportional to the displacement acting in the opposite direction of the displacement. The sum of potential and kinetic energies is the total energy in the mechanical system.

$$E = 0.5J\omega^2 + 0.5\frac{\tau^2}{K_r}$$

The sum of energies stored in the electric and magnetic fields is the total energy in the electrical analog.

$$E = 0.5Cv^2 + 0.5Li^2$$

Figure 4.18a and b show, respectively, the instantaneous energy stored in the electric and magnetic fields. At $t = 0$, the amplitude of the current is maximum at 10 and the reactive energy is $(0.5)(0.1)(10)(10) = 5$ var (volt-ampere reactive). As the voltage across the capacitor is zero, the reactive power stored in it is zero. As t increases, the current in the circuit charges the capacitor. At quarter of the cycle, the reactive power of the capacitor is maximum at $(0.5)(0.5)(\sqrt{20}\sqrt{20}) = 5$ var. The reactive

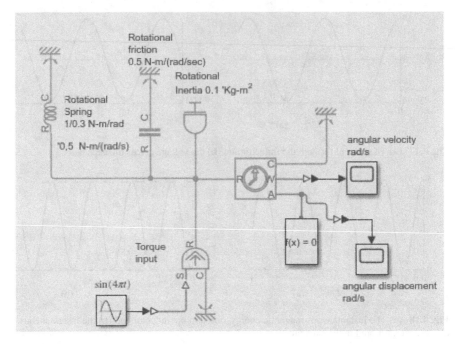

Fig. 4.19 A mechanical rotational system

power in the inductor is zero, as the current flowing through it is zero. The energy is swapped between magnetic and electric fields indefinitely resulting in oscillation. In the mechanical system, the kinetic and potential energies of the pendulum are swapped indefinitely resulting in oscillation. The total energy is a constant as shown in Fig. 4.18b by a dashed line. The period of the oscillation is determined by the length of the spring holding the mass and acceleration due to gravity. Similarly, the period of the oscillation in the equivalent electrical circuit depends on the values of the inductor and capacitor only.

4.3.2   A Mechanical Rotational System

Figure 4.19 shows a mechanical rotational system. Figure 4.20 shows the corresponding electrical analog to the system in Fig. 4.19. The input is current, $i(t) = \sin(4\pi t)$ A. The total admittance of the circuit is

$$Y_{tot}(s) = \frac{1}{R} + \frac{1}{sL} + sC = C \frac{\frac{s}{RC} + \frac{1}{LC} + s^2}{s}$$

The voltage $V(s)$ across the parallel circuit is

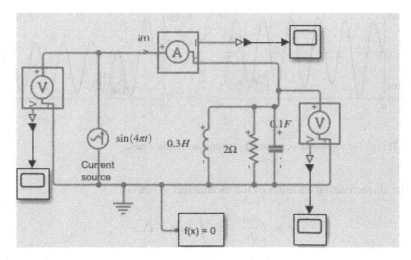

Fig. 4.20 The electrical analog the mechanical rotational system

$$V(s) = \frac{I(s)}{Y_{tot}(s)} = \frac{(4\pi/C)s}{(s^2 + (4\pi)^2)(\frac{s}{RC} + \frac{1}{LC} + s^2)}$$

System parameters are

$$C = 0.1F, \qquad R = 2\Omega, \qquad L = 0.3H$$

Substituting the numerical values, decomposing into partial fraction, and taking the inverse Laplace transform, we get

$$v(t) = 0.9992e^{-2.5t}\cos(5.2042t + 0.6354) + 0.9006\cos(4\pi t - 2.6745)$$

This voltage corresponds to angular velocity in the mechanical system. We have to integrate $v(t)$ to get the angular displacement of the mass $\theta(t)$. Integrating $v(t)$, we get

$$\theta(s) = \frac{(4\pi/C)}{(s^2 + (4\pi)^2)(\frac{s}{RC} + \frac{1}{LC} + s^2)}$$

Substituting the numerical values, decomposing into partial fraction, and taking the inverse Laplace transform, we get

$$\theta(t) = (0.1731e^{-2.5t}\cos(5.2042t - 1.3832) + 0.0717\cos(4\pi t + 2.0379))u(t)$$

Figure 4.21a and b shows the angular velocity and the angular displacement.

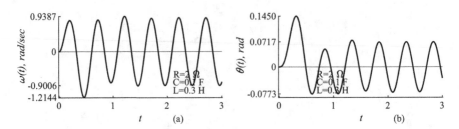

Fig. 4.21 (a) The angular velocity and (b) the angular displacement

The characterizing equation of the mechanical system is

$$\tau = J\frac{d\omega}{dt} + K_r \int \omega \, dt + B_r \omega$$

Using analogy, this equation can be written in the Laplace domain as

$$I_{in}(s) = V(s)\left(Cs + \frac{1}{Ls} + \frac{1}{R}\right) = V(s)\frac{RLCs^2 + R + sL}{RLs} = V(s)\frac{s^2 + \frac{1}{LC} + \frac{1}{RC}}{\frac{s}{C}}$$

This expression is the same as that obtained from the electrical analog with

$$I_{in}(s) = \frac{4\pi}{s^2 + (4\pi)^2}$$

4.3.3   Field Current Controlled DC Motor

In control systems, the device, called the actuator, provides motive power to the process. The DC motor is often the actuator and is a transducer converting electrical energy into rotational mechanical energy. One particular advantage is its speed controllability. As is the case in general, the model is developed, assuming linearity of the motor, as a compromise between the complexity and the accuracy of the model. In this example, we provide an example of simulating a system that spans electrical and mechanical domains. The DC motor is an electromechanical (a mechanical device that is operated by electricity) device. In deriving the transfer function of such devices, we write electrical and mechanical equations governing the device and use electromechanical relationships to couple the two sets.

Figure 4.22 shows the simulation model of the field-controlled DC motor generated by MATLAB® software. It has an electrical circuit on the left side and a mechanical circuit on the right side, connected by the rotational electromechanical converter block. This block is characterized by the equations

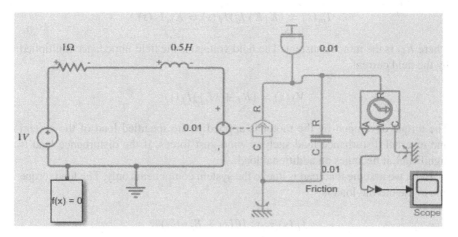

Fig. 4.22 Simulation model of the field-controlled DC motor

$$T = K_m I_f \quad \text{and} \quad V = K_m \omega$$

where T is the torque delivered to the load, ω is the angular velocity of the motor, I_f is the field current, and V is the applied voltage to the block. K_m is the motor constant, which is the same for the two sides with units N-m/A and V/(rad/s). The velocity of the motor is measured by the ideal rotational motion sensor and the scope. The load can be connected to the input terminal of the motion sensor. The rotational friction B_r, assumed of the viscous type, is also connected.

The basic principle on which the DC motor operation depends is that when a current carrying conductor is placed in a magnetic field, the conductor experiences a mechanical force. This results in the rotation of the armature. The armature is a coil in which voltage is induced due to its motion through the magnetic field. This induced electromotive force (emf) opposes the applied voltage $e_a(t)$ and is called back emf $e_b(t)$. The back emf regulates the armature current to meet the load requirements.

Let us derive the transfer function of the field current controlled DC Motor. The magnetic flux in the air gap of the motor is directly proportional to the field current $i_f(t)$,

$$\phi = K_f i_f(t)$$

Assuming linearity, the torque output T_m of the motor is

$$T_m = K_1 \phi i_a(t) = K_1 K_f i_f i_a(t)$$

In the field current controlled motor, the armature current $i_a(t)$ is held constant and the field current is the input variable. Maintaining one of the currents constant keeps the input output relationship linear. Taking the Laplace transform, we get

$$T_m(s) = (K_1 K_f I_a) I_f(s) = K_m I_f(s)$$

where K_m is the motor constant. The field voltage is the field impedance multiplied by the field current.

$$V_f(s) = (R_f + s L_f) I_f(s)$$

The torque developed by the motor is applied to the specified load of the system and external disturbance load such as wind-gust forces. If the disturbance load is significant, it becomes an additional load.

Here, we assume that load is due to the system components only. The load torque due to the inertial load is

$$T_L(s) = s^2 J \theta(s) + B_r s \theta(s)$$

and

$$T_m(s) = K_m I_f(s) = K_m \frac{V_f(s)}{(R_f + s L_f)}$$

From the last two identities, with $T_m(s) = T_L(s)$, we get the transfer function of the motor-load system as

$$\frac{\theta(s)}{V_f(s)} = H(s) = \frac{K_m}{s(sJ + B_r)(R_f + sL_f)} = \frac{K_m/(JL_f)}{s(s + B_r/J)(s + R_f/L_f)}$$

The time constant of the mechanical part of the system J/B_r is usually much longer than that of the electrical part L_f/R_f and may be neglected. The motor parameters, for example, are

$$R_f = 1\,\Omega, \quad L_f = 0.5\,\text{H}, \quad J = 0.01\,\text{Kg-m}^2,$$

$$B_r = 0.01\,\text{N-m/(rad/sec)}, \quad K_m = 0.01\,\text{V/(rad/sec)}$$

Substituting the parameters, the transfer function $H(s)$, with respect to ω, becomes

$$\frac{\omega(s)}{V_f(s)} = H(s) = \frac{2}{s^2 + 3s + 2}$$

Note that $\omega(s) = s\theta(s)$. With unit-step input voltage $u(t) \leftrightarrow 1/s$, the angular velocity of the motor is obtained by multiplying the input by the transfer function and taking the inverse Laplace transform

$$\omega(t) = (1 + e^{-2t} - 2e^{-t}) u(t)$$

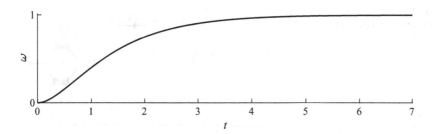

Fig. 4.23 Angular velocity versus time of the field-controlled DC motor

The angular velocity of the motor versus time is shown in Fig. 4.23. After the transient response, represented by exponential functions, becomes insignificant, the steady-state angular velocity is 1 rad/sec.

4.3.4 Armature-Controlled DC Motor

The block diagram representation of the armature-controlled DC motor is shown in Fig. 4.24. Let us derive the transfer function of the armature-controlled DC motor, in which the armature current i_a is the control variable and the field current remains fixed. With a fixed field current, the motor torque is

$$T_m(s) = (K_1 K_f I_f) I_a(s)$$

With the magnetic field generated by a permanent magnet,

$$T_m(s) = K_m I_a(s)$$

where K_m is a function of the permeability of the magnetic material. The back emf $V_b(s)$ is proportional to the angular speed $\omega(s)$ of the motor. Consequently,

$$V_b(s) = K_b \omega(s)$$

The armature current is given by

$$I_a(s) = \frac{V_a(s) - K_b \omega(s)}{R_a + s L_a}$$

Assuming the disturbance torque is 0, we got earlier

$$T_L(s) = s^2 J \theta(s) + B_r s \theta(s)$$

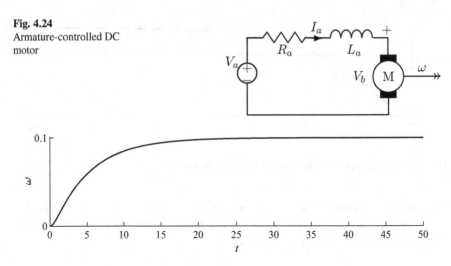

Fig. 4.24
Armature-controlled DC
motor

Fig. 4.25 Angular velocity versus time of the armature-controlled DC motor

The transfer function of the motor-load system, with $T_m(s) = T_L(s)$, is

$$\frac{\theta(s)}{V_a(s)} = H(s) = \frac{K_m}{s((sJ + B_r)(R_a + sL_a) + K_b K_m)}$$

The motor parameters, for example, are

$$R_f = 1\,\Omega,\ L_f = 0.5\ \text{H},\ J = 0.5\ \text{Kg-m}^2,\ B_r = 0.1\ \text{N-m/(rad/sec)},\ K_b = K_m = 0.01$$

Substituting the parameters, the transfer function $H(s)$, with respect to ω, becomes

$$\frac{\omega(s)}{V_f(s)} = H(s) = \frac{0.0400}{s^2 + 2.1995s + 0.4002}$$

With unit-step input voltage $u(t) \leftrightarrow 1/s$, the angular velocity of the motor is obtained by multiplying the input by the transfer function and taking the inverse Laplace transform

$$\omega(t) = (0.0999 + 0.0111e^{-1.9993t} - 0.1110e^{-0.2002t})u(t)$$

The angular velocity of the motor versus time is shown in Fig. 4.25. After the transient response, represented by exponential functions, becomes insignificant, the steady-state angular velocity is 0.0999 rad/sec.

4.4 Summary

- The basic components of mechanical systems are mass, spring, and damper. They are assumed to be ideal and linear in the operating range. With their characteristic equations and Newton's laws, we can model any linear mechanical system adequately.
- For translational systems, the algebraic sum of forces on a rigid body in a given direction is equal to the product of the mass of the body and its acceleration in the same direction is Newton's second law of motion.
- Physically different systems, such as mechanical and electrical systems, have similarity between their equilibrium equations. They are called analogous systems. This implies that electrical systems, which are easier to analyze, can be constructed whose behavioral characteristics are similar to the mechanical systems.
- A through variable is a variable that does not change between the ends of a circuit element, for example, the current flowing through resistor.
- An across variable is a variable that changes between the ends of a circuit element, for example, the voltage at the ends of a resistor.
- Topology of a network is the way the various components of a system are interconnected.
- One analogy preserves the topologies of the network of different systems those can be transformed with a one-to-one correspondence that is continuous in both directions. As voltage and velocity are across variables and current and force are through variables, the topology of the systems remains the same using this analogy.
- Translational mechanical systems (which move back and forth in a straight line) can be modeled using ideal masses, springs, and dampers.
- In rotational systems, mass is replaced by rotational mass or moment of inertia, linear spring is replaced by torsional spring, and damper is replaced by torsional damper.
- The algebraic sum of currents at a node is zero. Similarly, for mechanical systems, the algebraic sum of forces at an object is zero.
- For rotational systems, the algebraic sum of torques about a fixed axis is equal to the product of the inertia and the angular acceleration about the axis is the version of Newton's second law of motion.

Exercises

*4.1 Consider the frequency-domain representation of a resistor circuit with unit-step voltage source, shown in Fig. 4.26. Find the current $i(t)$ flowing through the circuit. Derive the equivalent expression for its translational mechanical analog and verify that the force through the mechanical system has the same numerical value as that of the current in the electrical circuit.

4.2 Consider the frequency-domain representation of a resistor circuit with unit-step current source, shown in Fig. 4.27. Find the voltage $v(t)$ across the

Fig. 4.26 Resistor circuit
with unit-step voltage source

Fig. 4.27 Resistor circuit
with unit-step current source

Fig. 4.28 Resistor–capacitor
circuit with unit-step voltage
source

Fig. 4.29 Resistor–inductor
circuit with unit-step voltage
source

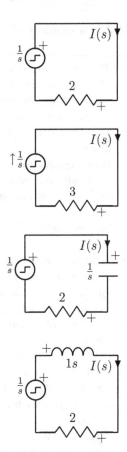

resistor. Derive the equivalent expression for its translational mechanical analog and verify that the velocity across damper has the same numerical value as that of the voltage in the electrical circuit.

*4.3 Consider the frequency-domain representation of a resistor–capacitor circuit with unit-step voltage source, shown in Fig. 4.28. Find the current $i(t)$ flowing through the circuit. Assume zero initial conditions. Derive the equivalent expression for its translational mechanical analog and verify that the force through the mechanical system has the same numerical value as that of the current in the electrical circuit.

4.4 Consider the frequency-domain representation of a resistor–inductor circuit with unit-step voltage source, shown in Fig. 4.29. Assume zero initial conditions. Find the current $i(t)$ flowing through the circuit. Derive the equivalent expression for its translational mechanical analog and verify that the force through the mechanical system has the same numerical value as that of the current in the electrical circuit.

Fig. 4.30
Resistor–inductor-capacitor
circuit with unit-step voltage
source

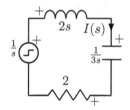

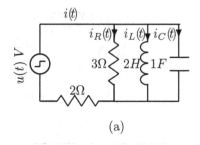

(a)

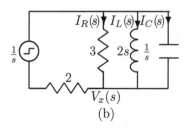

(b)

Fig. 4.31 Series–parallel circuit with a unit-step voltage source: (**a**) time-domain representation
and (**b**) frequency-domain representation

Fig. 4.32 Resistor–capacitor
circuit with unit-step voltage
source

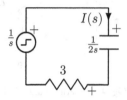

*4.5 Consider the frequency-domain representation of a resistor–inductor–
capacitor circuit with unit-step voltage source, shown in Fig. 4.30. Find
the current $i(t)$ flowing through the circuit. Assume zero initial conditions.
Derive the equivalent expression for its translational mechanical analog and
verify that the force through the mechanical system has the same numerical
value as that of the current in the electrical circuit.

4.6 Consider the frequency-domain representation of a series–parallel circuit with
unit-step voltage source, shown in Fig. 4.31. Find the current $i(t)$ flowing
through the circuit. Assume zero initial conditions. Derive the equivalent
expression for its translational mechanical analog and verify that the force
through the mechanical system has the same numerical value as that of the
current in the electrical circuit.

*4.7 Consider the frequency-domain representation of a resistor–capacitor circuit
with unit-step voltage source, shown in Fig. 4.32. Find the current $i(t)$
flowing through the circuit. Derive the equivalent expression for its rotational
mechanical analog and verify that the torque through the mechanical system
has the same numerical value as that of the current in the electrical circuit.

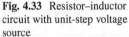

Fig. 4.33 Resistor–inductor circuit with unit-step voltage source

Fig. 4.34 Resistor–inductor-capacitor circuit with unit-step voltage source

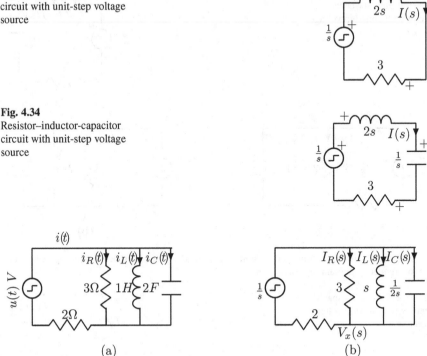

(a)   (b)

Fig. 4.35 Series–parallel circuit with a unit-step voltage source: (**a**) time-domain representation and (**b**) frequency-domain representation

4.8 Consider the frequency-domain representation of a resistor–inductor circuit with unit-step voltage source, shown in Fig. 4.33. Find the current $i(t)$ flowing through the circuit. Derive the equivalent expression for its rotational mechanical analog and verify that the torque through the mechanical system has the same numerical value as that of the current in the electrical circuit.

*4.9 Consider the frequency-domain representation of a resistor–inductor–capacitor circuit with unit-step voltage source, shown in Fig. 4.34. Find the current $i(t)$ flowing through the circuit. Assume zero initial conditions. Derive the equivalent expression for its rotational mechanical analog and verify that the torque through the mechanical system has the same numerical value as that of the current in the electrical circuit.

4.10 Consider the frequency-domain representation of a series–parallel circuit with unit-step voltage source, shown in Fig. 4.35. Find the current $i(t)$ flowing through the circuit. Assume zero initial conditions. Derive the equivalent expression for its rotational mechanical analog and verify that the torque through the mechanical system has the same numerical value as that of the current in the electrical circuit.

Chapter 5
Block Diagrams and Signal-Flow Graphs

Practical systems are composed of a very large number of subsystems. Therefore, analyzing the whole system at a time is almost impossible. However, by decomposing such systems into a suitable number of interconnected subsystems, each one of them can be analyzed relatively with much less effort. Each subsystem is characterized by its input–output relationship (for example, its transfer function). It is assumed that the loading effect of a subsystem connected to the output of another subsystem is negligible. **Block diagram** is a diagram showing the interconnection between the subsystems of a system. Some block diagrams of systems have been presented earlier. It is composed of blocks, signals, branch points, and summing junctions. Blocks represent subsystems, typically labeled with their transfer functions. Input and output signals of blocks are shown with their directions of flow indicated by arrows. Typical signals are voltage, force, temperature, pressure, and velocity, encountered in various applications. At summing junctions, incoming signals are algebraically summed or subtracted, subtraction being indicated by a minus sign. The signal at a point, called the branch point, is taken off to some other points.

If a linear system is excited by multiple inputs, it is easier to use the linearity property to find the response. At a time, in turn, set all inputs except one equal to zero and find the response. Then, reduce the block diagram with respect to the nonzero input. Repeat the procedure for each of the remaining inputs. The sum of all the responses is the response of the system with all the inputs acting simultaneously. Sometimes it is desirable to isolate a particular block to study the effect of it on the overall system. The usual reduction steps are used, but in a different order. The block to be isolated should not be combined with any others.

D. Sundararajan, *Control Systems*, https://doi.org/10.1007/978-3-030-98445-8_5

5.1 Block Diagrams

Consider the resistor–inductor circuit shown in Fig. 5.1. Due to voltage divider property of series circuits, the transfer function is

$$\frac{E_o(s)}{E_i(s)} = \frac{sL}{(R+Ls)}$$

The corresponding block diagram of the circuit is shown in Fig. 5.2. In feedback system theory, the transfer function of the unity feedback system is

$$\frac{E_o(s)}{E_i(s)} = \frac{sL/R}{1+sL/R} = \frac{sL}{(R+Ls)}$$

as obtained using circuit theory. Just to introduce the basics of block diagram theory, we used such a simple circuit, which does not require a block diagram representation. However, a block diagram approach is required to analyze high complexity circuits, such as that of a television receiver. The simplification of a block diagram is similar to simplifying an electrical circuit diagram. Rules for series and parallel connections are repeatedly used. Some equivalent circuits, such as $\Delta - Y$ transformation, are also used repeatedly. Of course, the input–output relationship of the system being simplified must remain the same after simplification.

Rules for simplifying block diagrams are as follows:

1. Combine all parallel blocks
2. Combine all cascaded blocks
3. Simplify all interior feedback loops
4. Shift branch points to the left or right if it helps simplification
5. Shift summing points to the left or right if it helps simplification

Fig. 5.1 A resistor–inductor circuit in Laplace domain

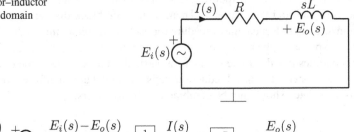

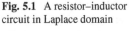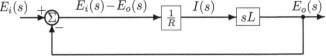

Fig. 5.2 Block diagram of a resistor–inductor circuit

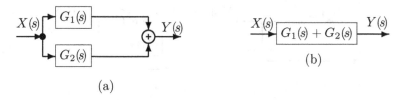

Fig. 5.3 (a) Blocks connected in parallel and (b) equivalent block

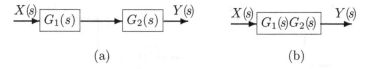

Fig. 5.4 (a) Blocks connected in cascade and (b) equivalent block

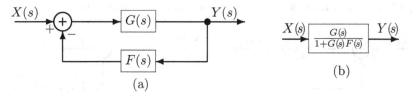

Fig. 5.5 (a) Blocks connected in a feedback configuration and (b) equivalent block

6. Simplify to a get a single block, using the above 5 rules repeatedly

Now, the simplification of blocks connected in three basic forms is presented. The equivalent transfer function $G_{eq}(s)$ of two systems connected in parallel is the sum of their individual transfer functions, $G_1(s)$ and $G_2(s)$, as shown in Fig. 5.3. That is,

$$G_{eq}(s) = G_1(s) + G_2(s)$$

The rule applies to any number of systems connected in parallel. The equivalent transfer function $G_{eq}(s)$ of two systems connected in cascade is the product of their individual transfer functions, $G_1(s)$ and $G_2(s)$, as shown in Fig. 5.4; that is,

$$G_{eq}(s) = G_1(s)G_2(s) = G_2(s)G_1(s)$$

The rule applies to any number of systems connected in cascade. Two systems connected in a feedback configuration and its equivalent representation are shown in Fig. 5.5.

Some more simplification of the block diagrams includes moving blocks forward/backward past summing junctions and branch points. Moving a branch point ahead of a block is shown in Fig. 5.6. Moving a branch point behind a block is shown

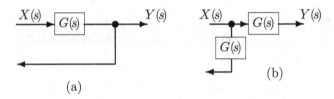

Fig. 5.6 (a) A block and (b) the result of moving a branch point ahead of the block

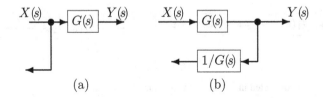

Fig. 5.7 (a) A block and (b) the result of moving a branch point behind the block

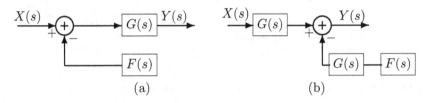

Fig. 5.8 (a) A block and (b) the result of moving a summing point behind the block

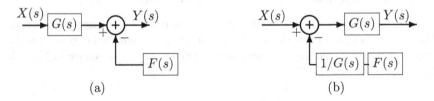

Fig. 5.9 (a) A block and (b) the result of moving a summing point ahead of the block

in Fig. 5.7. Moving a summing point behind a block is shown in Fig. 5.8. Moving a summing point ahead of a block is shown in Fig. 5.9.

Example 5.1 Consider the block diagram shown in Fig. 5.10. The two cascaded blocks $G_1(s)$ and $G_2(s)$ can be combined into a single block with transfer function $G_{12}(s) = G_1(s)G_2(s)$. The third summing point can be moved ahead. The two blocks connected in parallel $G_3(s)$ and $G_4(s)$ can be combined into a single block with transfer function $G_{34}(s) = G_3(s) + G_4(s)$, resulting in Fig. 5.11. Simplifying the feedback loop with $F_1(s)$ results in Fig. 5.12. Simplifying the top feedback loop results in Fig. 5.13. Simplifying the unity feedback loop results in Fig. 5.14. A block diagram with specific gain values is shown in Fig. 5.15.

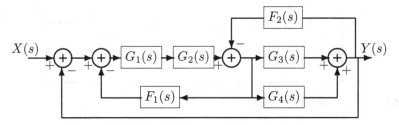

Fig. 5.10 A block diagram

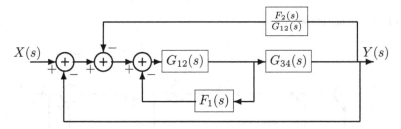

Fig. 5.11 Partially simplified block diagram

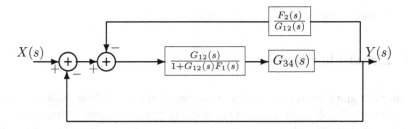

Fig. 5.12 Partially simplified block diagram

$$G_{12}(s) = G_1(s)G_2(s) = \frac{1}{s^2 + s}, \quad G_{34}(s) = G_3(s) + G_4(s) = \frac{3s + 2}{s},$$

$$F_1(s) = 0.1, \quad F_2(s) = 0.2$$

The transfer function is

$$H(s) = \frac{Y(s)}{X(s)} = \frac{G_{12}(s)G_{34}(s)}{1 + G_{12}(s)F_1(s) + G_{34}(s)F_2(s) + G_{12}(s)G_{34}(s)}$$

$$= \frac{\frac{1}{s^2+s}\frac{3s+2}{s}}{1 + 0.1\frac{1}{s^2+s} + 0.2\frac{3s+2}{s} + \frac{1}{s^2+s}\frac{3s+2}{s}} = \frac{3s + 2}{1.6s^3 + 2s^2 + 3.5s + 2}$$

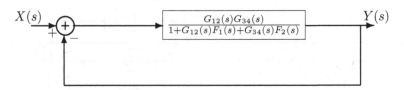

Fig. 5.13 Partially simplified block diagram

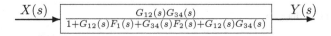

Fig. 5.14 Simplified block diagram

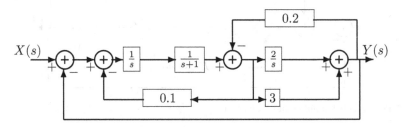

Fig. 5.15 A block diagram with specific gain values

5.2 Signal-Flow Graphs

An alternate graphical representation of systems is the signal-flow graphs (SFGs). SFG consists of nodes (representing signals) and branches (representing systems). Branches are labeled with transfer functions. The signal-flow direction is indicated by arrows. Summation is implied at the nodes. While both the block diagram and SFG give essentially the same information, one advantage of the SFG is that a gain formula, called Mason's gain formula, is available to find the relationships between variables. The SFG of a system is not unique. First, we should know the terminology used in describing SFG.

Node   A node shown by a disk represents a variable
Input node   An input node is a source of a variable
Output node   An output node is a sink of variables
Mixed node   A node that is a source as well as a sink
Gain   The gain between two nodes is indicated over an arrow pointing the direction of the signal flow.
Branch   A branch is a directed line segment connecting two nodes
Loop   A loop is a closed path. It starts at a node, passes through a set of nodes (passing through each node only once), and returns to the starting node
Loop gain   The loop gain is the product of the branch gains of a loop
Nontouching loops   Loops are nontouching if they do not share a common node

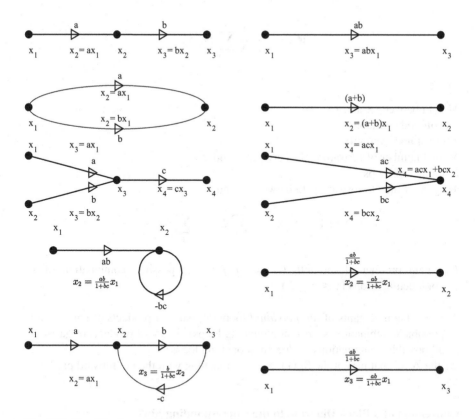

Fig. 5.16 Simplifications of SFG

Forward path   A path that starts at a source node and ends at a sink node that
does not cross any nodes more than once

Forward path gain   The gain that is the product of gains of a forward path

Some steps in the simplifications of SFG are shown in Fig. 5.16. The right side
figures are the simplified versions of those shown on the left. The last two pairs
show two versions of simplifications involving feedback loops. All these steps are
similar to those for block diagram reduction.

5.2.1 Mason's Gain Formula

The relationship between any two variables in an SFG can be obtained using
Mason's gain formula without reducing the SFG. For the determination of the input–
output relations, the gain formula is

$$M = \frac{y}{x} = \frac{1}{\Delta} \sum_{k=1}^{N} M_k \Delta_k$$

where

M gain between x and y
x input node variable
y output node variable
N total number of forward paths between x and y
L number of loops
M_k gain of the kth forward path between x and y

$$\Delta = 1 - \sum_i L_{i1} + \sum_j L_{j2} - \sum_k L_{k3} + \cdots$$

L_{mr} gain product of the mth ($m = i, j, k, \ldots$) possible combination of r nontouching loops ($1 \le r \le L$)
   or
$\Delta = 1-$ (sum of gains of *all individual* loops) + (sum of products of gains of all possible combinations of *two* nontouching loops) − sum of products of gains of all possible combinations of *three* nontouching loops) + $\cdots$
Δ_k the Δ for that part of the SFG that is nontouching with the kth forward path

Conversion of a Block Diagram to the Corresponding SFG

1. Place a node for each signal
2. Connect nodes with branches in place of the blocks
3. Indicate the correct direction with an arrow
4. Label branches with the corresponding transfer functions
5. Negate transfer functions to represent negative feedback

Find the transfer function of the system represented by the block diagram shown in Fig. 5.10 using Mason's formula.

Except that the variables are always summed at node points in SFG, there are no other differences between block diagram and SFG representation of a system. Therefore, the loop gains in the SFG must be made negative for each negative feedback loop in the block diagram representation. The gain of the two forward paths is

$$M_1 = G_1(s)G_2(s)G_3(s) \quad \text{and} \quad M_2 = G_1(s)G_2(s)G_4(s)$$

Individual Loops
There are five loops. The loop gains are

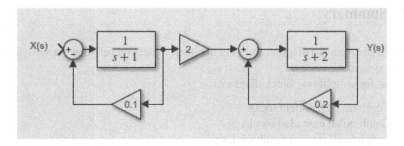

Fig. 5.17 Block diagram of a system

$$L_1 = -G_1(s)G_2(s)F_1(s), \qquad L_2 = -G_3(s)F_2(s), \qquad L_3 = -G_4(s)F_2(s),$$

$$L_4 = -G_1(s)G_2(s)G_3(s), \qquad L_5 = -G_1(s)G_2(s)G_4(s)$$

There are no nontouching loops.

$$\Delta = 1 - (L_1 + L_1 + L_3 + L_4 + L_5)$$
$$= 1 + G_1(s)G_2(s)F_1(s) + G_3(s)F_2(s) + G_4(s)F_2(s)$$
$$+ G_1(s)G_2(s)G_3(s) + G_1(s)G_2(s)G_4(s)$$

The transfer function, with $\Delta_1 = \Delta_2 = 1$, is

$$H(s) = \frac{M_1 + M_2}{\Delta} = \frac{G_1(s)G_2(s)G_3(s) + G_1(s)G_2(s)G_4(s)}{\Delta}$$

which is the same as that obtained using block diagram reduction.

Example 5.2 Find the transfer function of the block diagram, with input $X(s)$ and output $Y(s)$ shown in Fig. 5.17, by both block diagram reduction and Mason's formula. The transfer function of the system, from the block diagram, is

$$H(s) = \frac{Y(s)}{X(s)} = 2\frac{1}{s+1.1}\frac{1}{s+2.2} = \frac{2}{s^2 + 3.3s + 2.42}$$

From Mason's rule,

$$H(s) = \frac{Y(s)}{X(s)} = \frac{\frac{2}{(s+1)(s+2)}}{1 + \frac{0.1}{s+1} + \frac{0.2}{s+2} + \frac{0.1}{s+1}\frac{0.2}{s+2}} = \frac{2}{s^2 + 3.3s + 2.42}$$

There is one forward path and there are two individual loops and two nontouching loops.

5.3   Summary

- Block diagram is a diagram showing the interconnection between the subsystems of a system.
- Rules for simplifying block diagrams:

  1. Combine all parallel blocks
  2. Combine all cascaded blocks
  3. Simplify all interior feedback loops
  4. Shift branch points to the left or right if it helps simplification
  5. Shift summing points to the left or right if it helps simplification
  6. Simplify to get a single block, using the above 5 rules repeatedly

- An alternate graphical representation of systems is the signal-flow graphs. SFG consists of nodes (representing signals) and branches (representing systems). Branches are labeled with transfer functions. The signal-flow direction is indicated by arrows. Summation is implied at the nodes.
- The relationship between any two variables in an SFG can be obtained using Mason's gain formula without reducing the SFG.

Exercises

5.1 Consider the block diagram of a system with two inputs, shown in Fig. 5.18. Find the output $y(t)$, using the linearity property. Translate the block diagram into its equivalent signal-flow graph. Use Mason's gain formula also to find the transfer function of the system and verify that it is the same as that obtained by block diagram reduction.

*5.2 Consider the block diagram of a system with two inputs $x1(t)$ and $x2(t)$, shown in Fig. 5.19. Find the output $y(t)$, using the linearity property. Use Mason's gain formula also to find the transfer function of the system and verify that it is the same as that obtained by block diagram reduction.

5.3 Consider the block diagram of a system, shown in Fig. 5.20. Find the output $y(t)$ to the input $u(t)$. Use Mason's gain formula also to find the transfer function of the system and verify that it is the same as that obtained by block diagram reduction.

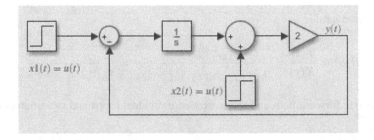

Fig. 5.18 Block diagram of a system with two inputs

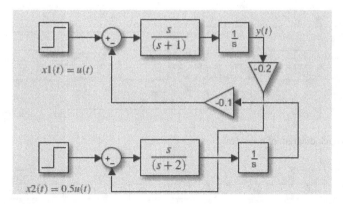

Fig. 5.19 Block diagram of a system with two inputs

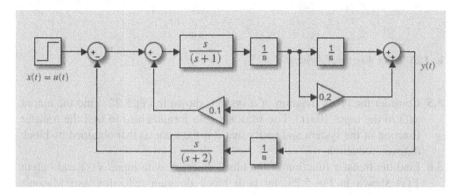

Fig. 5.20 Block diagram of a system

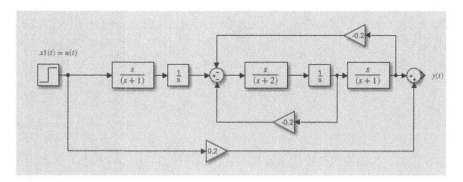

Fig. 5.21 Block diagram of a system

*5.4 Consider the block diagram of a system, shown in Fig. 5.21. Find the output
$y(t)$ to the input $u(t)$. Use Mason's gain formula also to find the transfer
function of the system and verify that it is the same as that obtained by block
diagram reduction.

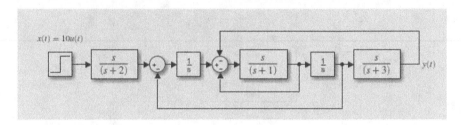

Fig. 5.22 Block diagram of a system

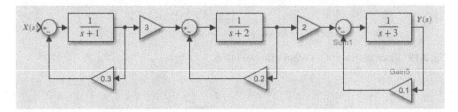

Fig. 5.23 Block diagram of a system

*5.5 Consider the block diagram of a system, shown in Fig. 5.22. Find the output $y(t)$ to the input $10u(t)$. Use Mason's gain formula also to find the transfer function of the system and verify that it is the same as that obtained by block diagram reduction.

5.6 Find the transfer function of the block diagram, with input $X(s)$ and output $Y(s)$ shown in Fig. 5.23, by both block diagram reduction and Mason's formula.

Chapter 6
Steady-State and Transient Responses

Most of the input signals to systems have arbitrary amplitude and it is difficult to characterize the response of the systems. Therefore, the response of the systems is determined for some standard signals, which is sufficient for the design and characterization of systems. The response to actual input signals can be determined from the response to standard signals by decomposing the input in terms of standard signals. When an input signal, such as the unit-step signal, is applied to a system, the output cannot follow the input instantaneously. This is so, due to the resistance offered by the system to the build-up of the output signal. For practical purposes, the duration of the resistance, called the transient interval or the interval of signal formation, lasts only for a short time for stable systems. The asymptotic behavior is called the steady-state or stationary or permanent state. A car cannot attain a high speed immediately after starting. Similarly, we cannot start running at high speed immediately after starting. In both cases, there is a build-up time. The total response of stable systems is composed of a transient component and a steady-state component.

6.1 Transfer Function of Feedback Systems

The transfer function of the feedback system is given as

$$H(s) = \frac{Y(s)}{X(s)} = \frac{G(s)}{1 + G(s)F(s)}$$

With unity feedback ($F(s) = 1$), $H(s)$ becomes

$$H(s) = \frac{Y(s)}{X(s)} = \frac{G(s)}{1 + G(s)}$$

The transfer function between the error signal $E(s)$ and the input signal $X(s)$ is

$$\frac{E(s)}{X(s)} = 1 - \frac{Y(s)}{X(s)} = \frac{1}{1 + G(s)}$$

The Laplace transform of the error function is

$$E(s) = \frac{X(s)}{1 + G(s)}$$

6.2   Steady-State Errors in Control Systems

For stable systems, the transient part of the total response tends to zero. After it dies down to negligible levels (typically after 4 time constants of the system for practical systems), what remains is the steady-state response. However, the actual steady-state response may not be the same as expected. The difference between the expected and actual responses is the steady-state error. The steady-state error depends on the input and the type of the system. The error is, using the final value theorem,

$$e(\infty) = e_{ss} = \lim_{s \to 0} s E(s) = \lim_{s \to 0} \frac{s X(s)}{1 + G(s)}$$

With the input $X(s)$ being the unit-step signal $u(t) \leftrightarrow 1/s$, the error is

$$e(\infty) = \frac{1}{1 + \lim_{s \to 0} G(s)} = \frac{1}{1 + K_p}, \quad \text{with } K_p = \lim_{s \to 0} G(s)$$

With unit-ramp input $r(t) \leftrightarrow 1/s^2$, the error is

$$e(\infty) = \lim_{s \to 0} \frac{1}{s + s G(s)} = \frac{1}{K_v}, \quad \text{with } K_v = \lim_{s \to 0} s G(s)$$

With parabolic input $p(t) \leftrightarrow 1/s^3$, the error is

$$e(\infty) = \frac{1}{K_a}, \quad \text{with } K_a = \lim_{s \to 0} s^2 G(s)$$

The static error constants defined are figures of merit for control systems. The higher the value of the constants, the smaller is the steady-state error. These constants are called position, velocity, and acceleration error constants. For different physical systems, these error constants must be identified accordingly. For example, in a temperature control system, the position error corresponds to the change in the output temperature. Knowing the constants K_p, K_v and K_a, and the system type, we can determine whether the system has a finite steady-state error. **System type** is the number of pure integrators (poles at $s = 0$) the system has in its forward path of a unity feedback system.

6.2.1 Type 0 System

Consider the loop transfer function of a Type 0 unity feedback system

$$L(s) = K\frac{s+1}{s+2}$$

with $K = 1$. The position error constant is

$$K_p = \lim_{s\to0}\frac{s+1}{s+2} = \frac{1}{2}$$

The steady-state error is

$$e(\infty) = e_{ss} = \frac{1}{1+0.5} = \frac{2}{3}$$

as shown by the dashed line in Fig. 6.1a. For understanding and verification, we have also found the response of the unity feedback system directly. This figure shows the response $y(t)$ of the corresponding closed-loop system with the transfer function

$$H(s) = \frac{L(s)}{1+L(s)} = \frac{s+1}{2s+3}$$

to the unit-step input $u(t)$ by the solid line. Multiplying the transfer function by the unit-step input $u(t) \leftrightarrow 1/s$, and taking the inverse Laplace transform, the output is

$$y(t) = \left(\frac{1}{3} + \frac{1}{6}e^{-1.5t}\right)u(t)$$

As $t \to \infty$, the response tends to $1/3$, making the error $2/3$. The error is the same as that obtained using $L(s)$.

Consider

$$L(s) = K\frac{(s+1)}{s+2}$$

with $K = 4$. The position error constant is

$$K_p = \lim_{s\to0}\frac{4(s+1)}{s+2} = 2$$

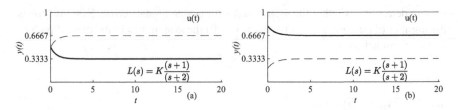

Fig. 6.1 (a) Steady-state error with $K = 1$; (b) Steady-state error with $K = 4$

The steady-state error is

$$e(\infty) = \frac{1}{1+2} = \frac{1}{3}$$

The response becomes

$$y(t) = \left(\frac{2}{3} + \frac{2}{15}e^{-1.2t}\right)u(t)$$

As $t \rightarrow \infty$, the response tends to 2/3, making the error 1/3, as shown in Fig. 6.1b. The error is the same as that could be obtained using $L(s)$. As the gain is increased, the error is reduced. However, increasing gain aggravates the stability problem. A compromise between the relative stability and steady-state accuracy is required. For zero steady-state error, a Type 1 or higher system is required.

6.2.2   Type 1 System

Consider the loop transfer function of a Type 1 unity feedback system

$$L(s) = K\frac{s+1}{s(s+2)}$$

with $K = 1$. The velocity error constant is

$$K_v = \lim_{s \to 0} s \frac{s+1}{s(s+2)} = \frac{1}{2}$$

The steady-state error is

$$e(\infty) = \frac{1}{0.5} = 2$$

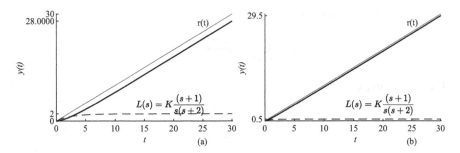

Fig. 6.2 (a) Steady-state error with $K = 1$; (b) Steady-state error with $K = 4$

as shown in Fig. 6.2a. For understanding and verification, we have also found the response of the unity feedback system directly. This figure shows the response $y(t)$ of the corresponding closed-loop system with the transfer function

$$\frac{L(s)}{1 + L(s)} = \frac{s + 1}{s^2 + 3s + 1}$$

to a unit-ramp input $r(t)$. Multiplying the transfer function by the unit-ramp input $r(t) \leftrightarrow 1/s^2$, and taking the inverse Laplace transform, the output is

$$y(t) = (t - 2 + 1.8944e^{-0.3820t} + 0.1056e^{-2.6180t})u(t)$$

As $t \to \infty$, the response tends to $t - 2$, making the error 2. The error is the same as that could be obtained using $L(s)$.

Consider

$$H(s) = K\frac{(s + 1)}{s(s + 2)}$$

with $K = 4$. The response becomes

$$y(t) = (t - 0.5 + 0.3618e^{-0.7639t} + 0.1382e^{-5.2361t})u(t)$$

As $t \to \infty$, the response tends to $t - 0.5$, making the error 0.5, as shown in Fig. 6.2b. The error is the same as that could be obtained using $L(s)$.

6.2.3 Type 2 System

Consider the loop transfer function of a Type 2 unity feedback system

$$L(s) = K\frac{s + 2}{s^2(s + 3)}$$

with $K = 1$. The acceleration error constant is

$$K_a = \lim_{s \to 0} s^2 \frac{s+2}{s^2(s+3)} = \frac{2}{3}$$

The steady-state error is

$$e(\infty) = \frac{3}{2} = 1.5$$

For understanding and verification, we have also found the response of the system directly. The closed-loop transfer function is

$$\frac{s+2}{s^3 + 3s^2 + s + 2}$$

Multiplying the transfer function by the unit-parabolic input $p(t) \leftrightarrow 1/s^3$, and taking the inverse Laplace transform, the output is

$$y(t) = (1.4999e^{-0.0534t} \cos(0.8297t - 0.0739) + 0.0042e^{-2.8933t} + 0.5t^2 - 1.5)u(t)$$

As $t \to \infty$, the response tends to $0.5t^2 - 1.5$, making the error 1.5. The error is the same as that could be obtained using $L(s)$. With $K = 4$, the response becomes

$$y(t) = (0.3648e^{-0.1828t} \cos(1.7330t - 0.1690) + 0.0154e^{-2.6344t} + 0.5t^2 - 0.3750)u(t)$$

As $t \to \infty$, the response tends to $0.5t^2 - 0.375$, making the error 0.375. The error is the same as that could be obtained using $L(s)$. These examples show that getting the steady-state error is much easier from the open-loop transfer function rather than from the closed-loop transfer function. Such is the case, in general, in the analysis and design of control systems. The steady-state errors for three types of systems with three standard inputs are shown in Table 6.1.

Table 6.1 Steady-state errors of unity feedback systems

System	Step input $x(t) = u(t) = 1$	Ramp input $x(t) = r(t) = t$	Acceleration input $x(t) = 0.5t^2$
Type 0	$e_{ss} = \frac{1}{1+K_p}$	∞	∞
Type 1	$e_{ss} = 0$	$\frac{1}{K_v}$	∞
Type 2	$e_{ss} = 0$	0	$\frac{1}{K_a}$

Steady-State Errors of Nonunity Feedback Systems

As the error constants are defined using the forward path transfer function $G(s)$, the definitions are applicable to unity feedback systems only. Consider the transfer function of the nonunity feedback system.

$$H(s) = \frac{Y(s)}{X(s)} = \frac{G(s)}{1 + G(s)F(s)}$$

For steady-state error analysis, this system can be converted to an equivalent unity feedback system with

$$L(s) = \frac{G(s)}{1 + G(s)(F(s) - 1)}$$

As the error constants are defined using the final value theorem of the Laplace transform, the ROC of $sL(s)$ must include the $j\omega$ axis.

6.3   Unit-Step Response and Transient Response Specifications

In practice, the order of the control system is very high. But, for implementation advantages, they are usually decomposed into first- and second-order systems. Therefore, the analysis of first- and second-order systems is necessary for understanding the analysis and design of higher-order systems. The impulse response of stable systems does not have the steady-state component. Consequently, the unit-step signal is widely used as the test signal to analyze and design control systems.

Consider the open-loop transfer function of a second-order system with unity feedback in standard form.

$$G(s) = \frac{Y(s)}{E(s)} = \frac{\omega_n^2}{s^2 + 2\zeta\omega_n s}$$

where ζ is the damping ratio and ω_n is the undamped natural frequency in rad/sec. The corresponding closed-loop transfer function is

$$H(s) = \frac{Y(s)}{X(s)} = \frac{\omega_n^2}{s^2 + 2\zeta\omega_n s + \omega_n^2}$$

This is just a second-order Laplace transfer function, but the coefficients are expressed in terms of important system parameters. This form is called the standard form of the second-order transfer function.

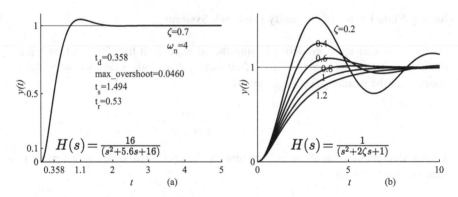

Fig. 6.3 (a) Unit-step response; (b) unit-step response with various damping ratios ζ

Substituting the specific values $\zeta = 0.7$ and $\omega_n = 4$ rad/sec, we get

$$H(s) = \frac{16}{s^2 + 5.6s + 16}$$

Multiplying by the unit-step function $u(t) \leftrightarrow 1/s$, decomposing into partial fraction and taking the inverse Laplace transform, we get the unit-step response as

$$y(t) = 1 + 1.4003e^{-2.8t} \cos(2.8566t + 2.3662)$$

The unit-impulse response is obtained by differentiating the unit-step response. Figure 6.3a shows the unit-step response of the second-order transfer function. The characteristic figures are shown in the figure. As both the transient and steady-state responses are critical for control systems, these specifications are quite important. In most systems, typically, the damping ratio is between 0.4 and 0.8 to avoid excessive overshoot and sluggish response. Note that, maximum overshoot and rise time conflict each other.

Figure 6.3b shows the unit-step response of the second-order transfer function

$$H(s) = \frac{1}{(s^2 + 2\zeta s + 1)}$$

with $\omega_n = 1$ rad/sec and various values of ζ. The unit-step response is usually measured with zero initial conditions, which makes it easy to compare with those of other systems.

The response is of three types. If $\zeta = 1$, both the roots are real and the same (critically damped). If $\zeta < 1$, the roots are complex-conjugates (underdamped) with negative real parts. If $\zeta > 1$, both the roots are real (overdamped). For the underdamped case $0 < \zeta < 1$, the unit-step response is given as

$$y(t) = (1 + \frac{e^{-\zeta \omega_n t}}{\sqrt{(1 - \zeta^2)}} \cos(\omega_n \sqrt{(1 - \zeta^2)}t + (\cos^{-1}(\zeta) + \frac{\pi}{2})))u(t)$$

The **damped frequency** is $\omega_d = \omega_n\sqrt{1 - \zeta^2}$ and usually less than ω_n. Substituting $\zeta = 0.7$ and $\omega_n = 4$ rad/sec in the general expression given above, we get the specific response for $y(t)$ shown in Fig. 6.3a. The damped frequency is 2.8566 rad/sec, as can be seen from the expression for $y(t)$. The two roots of the underdamped ($\zeta < 1$) second-order transfer function in standard form, using the quadratic formula, are

$$-\zeta\omega_n \pm j\omega_n\sqrt{1 - \zeta^2} = -\alpha \pm j\omega_d$$

with $\alpha = \zeta\omega_n$ and $\omega_d = \omega_n\sqrt{1 - \zeta^2}$. The response becomes more oscillatory for lower values of ζ. For the overdamped case, the roots are

$$-\zeta\omega_n \pm \omega_n\sqrt{\zeta^2 - 1}$$

For the critically damped case $\zeta = 1$, the roots are $-\omega_n$. For $\zeta \geq 1$, the response never exceeds its final value. For the undamped case $\zeta = 0$, the roots are $\pm j\omega_n$. In this case, the response is a steady sinusoid. In the expression for unit-step response, α appears in the exponential term, Therefore, it controls the rise or decay of the response.

The maximum overshoot is

$$M_p = e^{-\frac{\pi\zeta}{\sqrt{1-\zeta^2}}}$$

and it occurs at

$$t_p = \frac{\pi}{\omega_n\sqrt{1 - \zeta^2}}$$

The **maximum overshoot** of the response $y(t)$ is defined as the difference between the maximum value of $y(t)$ and the steady-state value $\lim_{t\to\infty} y(t)$. The **peak time** t_p is the time at which the first peak of the response occurs. For the example, the value is 0.046 at $t = 1.1$ sec. (or 4.6%) as shown in the figure.

The **settling time** t_s is defined as the time required for the response to be within certain percentage of its final value, typically 2%. The settling time t_s has to satisfy the condition

$$e^{-\zeta\omega_n t_s} < 0.02$$

That is,

$$e^{\zeta\omega_n t_s} = \frac{1}{0.02} = 50 \quad \text{or} \quad t_s = \frac{\log_e(50)}{\zeta\omega_n} = \frac{3.9120}{\zeta\omega_n} \cong \frac{4}{\zeta\omega_n} = 4\tau,$$

where τ is the time constant. Consider only the exponential part of the unit-step response

$$1.4003e^{-2.8t}u(t) = 1.4003e^{-\frac{t}{0.3571}}u(t)$$

At $t = 0.3571$, the response becomes

$$1.4003e^{-1}u(t) = 1.4003(0.3679) = 0.5152$$

That is, the value of the exponential decreases to 0.3679 of its initial value in one time constant $\tau = 0.3571$ seconds. In 4 time constants, the value decreases to $0.3679^4 = 0.0183$ which is less than 2% of its initial value. We approximated this to 2% to define the settling time. The system with a smaller time constant is faster that responds quickly to the input.

The **delay time** t_d is the time required for the response to reach 50% of its final value for the first time. The **rise time** is the time required for the response to rise from 10% to 90% (usually used for overdamped systems) or 0% to 100% (usually used for underdamped systems) of its final value.

Damping Ratio and Damping Factor
For the case of critical damping with $\zeta = 1$, the magnitude of the real part of the roots of the characteristics of the equation of the system are the same, ω_n with the imaginary part zero. The factor $\zeta \omega_n$, called the damping factor, actually controls the damping of the system. Then, ζ can be considered as the damping ratio

$$\zeta = \frac{\zeta \omega_n}{\omega_n} = \frac{\text{actual damping factor}}{\text{damping factor at critical damping}}$$

6.4 Linearization

There are abundant analytical tools to analyze linear systems. However, most practical systems are nonlinear to some extent. Such systems are approximated to be linear within some range and analyzed as linear systems. One of the often used example is the value of $\sin(\theta) \cong \theta$ as $\theta \to 0$. The Maclaurin series expansion for the sine function is

$$\sin(\theta) = \theta - \frac{\theta^3}{3!} + \frac{\theta^5}{5!} - \cdots + (-1)^r \frac{\theta^{2r+1}}{(2r+1)!} - \cdots$$

Now, if we retain only the linear term, the definition becomes linear. It is assumed that the contribution of the neglected terms to the sum is negligible. An example is the linearization of the simple pendulum model. Further, $\sin(\theta) \cong 1$ as $\theta \to \frac{\pi}{2}$. The linear approximation of functions $f(x)$ can be obtained by taking the first two terms of their Taylor series expansions about a point x_0.

$$f(x) = f(x_0) + \frac{d}{dx} f(x)|_{x=x_0} (x - x_0)$$

A nonlinear function can be represented by a set of linear functions defined at several points. Another way to analyze nonlinear functions is by simulation.

6.5 Parameter Sensitivity

Due to changes in environmental conditions and aging, the behavior of physical components tends to deviate from the ideal one. Consequently, the system performance changes from that expected. The variation of the response of systems is characterized by the sensitivity function. Typical solutions to reduce the sensitivity are the use of high-quality components, better structures for implementation and decomposing higher-order transfer functions into a set of lower-order transfer functions.

Consider the input–output relationship of a system with the open-loop transfer function $G(s)$ in the s-domain

$$Y(s) = G(s)X(s)$$

where $X(s)$ is the input and $Y(s)$ is the output. For a given input function, differentiating the output expression with respect to $G(s)$ yields the effect of a change in $G(s)$. That is,

$$dY(s) = dG(s)X(s) = \frac{dG(s)}{G(s)}Y(s)$$

and the sensitivity function is given by

$$S_{G(s)}^{Y(s)}(s) = \frac{\frac{dY(s)}{Y(s)}}{\frac{dG(s)}{G(s)}} = 1$$

A change in $G(s)$ causes a proportional change in $Y(s)$.

Let $G(s) = \frac{1}{s+1}$ and the input $u(t)$, the unit-step function. Then, the output is

$$Y(s) = \frac{1}{s(s+1)}$$

Let the changed transfer function be $G(s) = \frac{1}{s+1.01}$. Then, the output is

$$Y(s) = \frac{1}{s(s+1.01)}$$

The steady-state response of the system to unit-step input is 0.9054 rather than 1 due to sensitivity, as shown in Fig. 6.4.

Consider the input–output relationship of the corresponding closed-loop transfer function

$$Y(s) = \left(\frac{G(s)}{1 + G(s)}\right) X(s)$$

Differentiating this expression with respect to $G(s)$ yields

$$dY(s) = \frac{dG(s)}{(1 + G(s))^2} X(s) = \frac{1}{(1 + G(s))} \frac{dG(s)}{G(s)} Y(s)$$

and the sensitivity function is given by

$$S_{G(s)}^{Y(s)}(s) = \frac{\frac{dY(s)}{Y(s)}}{\frac{dG(s)}{G(s)}} = \frac{1}{(1 + G(s))}$$

Sensitivity is reduced by the factor $\frac{1}{(1+G(s))}$ compared with the open-loop system. Let $G(s) = \frac{1}{s+1}$. Then, the output of the closed-loop system is

$$Y(s) = \frac{1}{s(s + 2)}$$

Let the changed transfer function be $G(s) = \frac{1}{s+1.01}$. Then, the output is

$$Y(s) = \frac{1}{s(s + 2.01)}$$

The steady-state response of the system to unit-step input is 0.4762 rather than 0.5 due to sensitivity, as shown in Fig. 6.4. The relative error due to sensitivity is reduced by about a factor of 2, as expected. In addition to providing the desired output with a specified accuracy, stable closed-loop feedback systems has other advantages also. For a nonunity feedback system, the sensitivity factor is better with

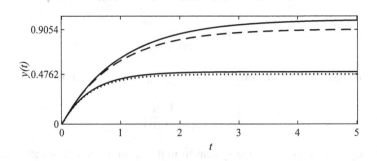

Fig. 6.4 Variation of step response due to sensitivity

$$\frac{1}{(1 + G(s)F(s))}$$

Using feedback compensation, the response of the system can be made to depend on the feedback path gain rather than the forward path gain.

6.6 Summary

- The response of the systems is determined for some standard signals, which is sufficient for the design and characterization of systems.
- The total response of stable systems is composed of a transient component and a steady-state component.
- For stable systems, the transient part of the total response tends to negligible levels (typically after 4 time constants of the system for practical systems). Afterward, the response is the steady-state response.
- The actual steady-state response may not be the same as expected. The difference between the expected and actual responses is the steady-state error.
- The steady-state error depends on the input and the type of the system. The error is, using the final value theorem,

$$e(\infty) = e_{ss} = \lim_{s \to 0} s E(s) = \lim_{s \to 0} \frac{s X(s)}{1 + G(s)}$$

- The static error constants defined are figures of merit for control systems. The higher the value of the constants, the smaller is the steady-state error.
- System type is the number of pure integrators (poles at $s = 0$) the system has in its forward path of a unity feedback system.
- As the error constants are defined using the forward path transfer function $G(s)$, the definitions are applicable to unity feedback systems only.
- The impulse response of stable systems does not have the steady-state component. Consequently, the unit-step signal is widely used as the test signal to analyze and design control systems.
- The closed-loop transfer function of second-order systems, in standard form, is

$$H(s) = \frac{Y(s)}{X(s)} = \frac{\omega_n^2}{s^2 + 2\zeta \omega_n s + \omega_n^2}$$

where ζ is the damping ratio and ω_n is the undamped natural frequency in rad/sec. The coefficients of the transfer function are expressed in terms of important system parameters.

- The important parameters characterizing the transient response are maximum overshoot, settling time, rise time, delay time, and peak time.

- Most practical systems, which are nonlinear to some extent, are approximated to be linear within some range and analyzed as linear systems for convenience.
- Due to changes in environmental conditions and aging, the behavior of physical components tends to deviate from the ideal one. Consequently, the system performance changes from that expected. The variation of the response of systems is characterized by the sensitivity function.

Exercises

6.1 Consider the loop transfer function $L(s)$ of a Type 0 unity feedback system. Find the steady-state error between the unit-step input $u(t)$ and the output $y(t)$. Find also an expression for the total output $y(t)$ and verify that the steady-state error is the same.

*6.1.1

$$L(s) = \frac{1}{s+2}$$

6.1.2

$$L(s) = \frac{s+2}{s+1}$$

6.1.3

$$L(s) = \frac{s+2}{(s+1)(s+3)}$$

6.2 Consider the loop transfer function $L(s)$ of a Type 1 unity feedback system. Find the steady-state error between the unit-ramp input $r(t)$ and the output $y(t)$. Find also an expression for the total output $y(t)$ and verify that the steady-state error is the same.

6.2.1

$$L(s) = \frac{1}{s(s+2)}$$

*6.2.2

$$L(s) = \frac{s+2}{s(s+1)}$$

6.2.3

$$L(s) = \frac{s+2}{s(s+1)(s+3)}$$

6.3 Consider the loop transfer function $L(s)$ of a Type 2 unity feedback system. Find the steady-state error between the parabolic input $0.5t^2u(t)$ and the output $y(t)$. Find also an expression for the total output $y(t)$ and verify that the steady-state error is the same.

6.3.1

$$L(s) = \frac{s+1}{s^2(s+2)}$$

6.3.2

$$L(s) = \frac{s^2 + 2s + 1}{s^2(s^2 + 5s + 6)}$$

*6.3.3

$$L(s) = \frac{s^2 + 2s + 1}{s^2(s^2 + 6s + 8)}$$

6.4 Consider the closed-loop second-order transfer function $H(s)$ of a feedback system. Find $\omega_n, \omega_d, \zeta, t_p, M_p, t_s$ for the system using the definitions for the unit-step input $u(t)$. Find also an expression for the output $y(t)$, measure the above mentioned parameters from the response and verify that the values are the same.

*6.4.1

$$H(s) = \frac{16}{s^2 + 4s + 16}$$

6.4.2

$$H(s) = \frac{4}{s^2 + 2.8s + 4}$$

6.4.3

$$H(s) = \frac{1}{s^2 + 1.2s + 1}$$

Chapter 7
Root Locus

Basically, a physical system is adequately modeled by differential equations using the interconnection of the components and laws governing the behavior of each component. By varying some key parameters of the system, a system can be designed to meet the specifications by repeatedly solving the differential equations. But, this method is very tedious and impractical for practical systems. There are three major tools used to easily analyze and design control systems, the root locus, Bode, and Nyquist plots. These tools bring out from the model some important characteristics of the system those can be used to analyze and design systems easily. We can get used to these tools with sufficient manual (for simpler systems) and programming practice. Using these tools for the design of control systems is similar to using a table of logarithms to simplify the multiplication operation.

Both the transient and steady-state responses of the system are important in control system analysis. The transient response is composed of exponentials of the form e^{-pt}, where t is the independent variable and p are the characteristic modes (the roots of the characteristic equation of the closed-loop transfer function) of the system. The roots are the poles of the system, which vary with the value of the gain K. Assume that K is positive. The system is asymptotically stable only if its poles are located in the left-half of the s-plane. That is, the real part of the root must be negative. The gain can be increased until the roots are located in the right-half of the s-plane. As the roots primarily determine the transient response, it helps to know the variations in the location of the poles as the system gain is varied. The gain adjustment along with a compensator is required to get a desirable response from a feedback control system.

D. Sundararajan, *Control Systems*, https://doi.org/10.1007/978-3-030-98445-8_7

7.1   Plotting the Root Locus

7.1.1   Negative Feedback Systems

Consider a negative feedback system with feed-forward transfer function $G(s)$ and the feedback transfer function $F(s)$. The closed-loop transfer function is

$$H(s) = \frac{G(s)}{1 + G(s)F(s)}$$

The loop transfer function is $G(s)F(s)$. Often, $G(s)F(s)$ is a rational function of the form

$$\frac{N(s)}{D(s)}$$

where $N(s)$ and $D(s)$ are the numerator and denominator polynomials in s. In the case of systems with pure delay, $G(s)F(s)$ can be approximated by a rational function of polynomials. The **root locus** of a system with characteristic equation $1 + G(s)F(s) = 0$ is the set of loci of the closed-loop poles as some parameter varies (typically, as the proportional gain varies from 0 to infinity). The poles are the roots of the characteristic equation

$$1 + G(s)F(s) = 0$$

The basis of control system design, based on the root locus and other tools, is that the behavior of the system can be modified in a desired way by proper placement of its poles and zeros. A pole and a zero nearby tend to cancel out each other's influence. By adding open-loop poles and zeros to $G(s)F(s)$, the root locus can be shifted to the right or left to suit the requirements. The root locus is symmetrical about the real axis, since complex roots occur in conjugate pairs for practical systems.

Let $G(s)F(s)$, in terms of its numerator and denominator polynomials, be

$$G(s)F(s) = \frac{Kb(s)}{a(s)}$$

with K being a variable parameter. The numerator and denominator coefficients are usually designated, respectively, using the letters b and a. The coefficients are assumed to be real and fixed. Then, the characteristic equation becomes

$$a(s) + Kb(s) = 0$$

As $K \to 0$, the characteristic equation reduces to

$$a(s) = 0$$

Dividing the characteristic equation by K, we get

$$\frac{a(s)}{K} + b(s) = 0$$

As $K \to \infty$, the characteristic equation reduces to

$$b(s) = 0$$

Therefore, the root locus starting points are at the loop transfer function poles with $K = 0$. The root locus ending points are at the loop transfer function zeros with $K = \infty$.

Irrespective of the value of K, the number of poles of the loop and closed-loop transfer functions are the same. There are the same number of branches in the root locus, with each branch starting at a pole and ending at a zero of the loop transfer function. Let n be the number of poles and m be the number of zeros. Usually, $m < n$. Then, there are $n - m$ zeros at infinity and, consequently, $n - m$ branches of the locus end at infinity. The remaining branches terminate at finite zeros.

The characteristic equation is

$$1 + G(s)F(s) = 0 \quad \text{or} \quad G(s)F(s) = -1$$

As $G(s)F(s)$ is a complex quantity, the condition $G(s)F(s) = -1 = e^{j(1+2k)\pi}$ can be split into magnitude and phase angle conditions. The phase angle condition is

$$\angle(G(s)F(s)) = \pm 180°(2k + 1), \quad k = 0, 1, 2, \ldots$$

The magnitude condition is

$$|G(s)F(s)| = 1$$

A plot of the points in the complex plane, in terms of real and imaginary parts of $s = \sigma + j\omega$, satisfying the angle condition alone is an alternative definition of the **root locus**. Once the root loci is drawn, the values of K on the loci can be determined as

$$K = \frac{1}{|G(s)H(s)|}$$

Example 7.1.1 Figure 7.1a shows the root locus of the loop transfer function $\frac{K}{s+1}$ with the variable parameter K. The location of the only pole of the loop transfer function is indicated by a cross. The corresponding characteristic equation of the closed-loop transfer function is $s + (1 + K)$. With $K = 0$, the root is -1. As K is increased, the root becomes $-(K + 1)$. The arrow indicates the direction of

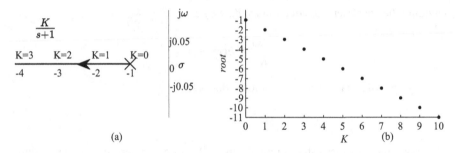

Fig. 7.1 (a) Root locus of $\frac{K}{s+1}$; (b) gain K versus root

increasing gain. Figure 7.1b shows the root versus K plot. The magnitude of the gain K at $s = s_0$ is $|s_0 + 1|$.

$$\text{At } s_0 = -1, K = |-1+1| = 0, \text{ at } s_0 = -2, K = |-2+1| = 1,$$
$$\text{at } s_0 = -3, K = |-3+1| = 2$$

as shown in Fig. 7.1a and b. In this simple example, the variation of the root with K is obvious. The **number of individual root loci** is the same as the number of poles of the loop transfer function. There is only one branch in the plot, as the number of poles is one. The root locus starts at -1, the pole of the loop transfer function, and moves toward $-\infty$. As there is no finite zero, the locus terminates at $-\infty$.

Rule 1 Root loci on the real axis   At a given point of the real axis, the root locus exists only if the sum of the number of all the real poles and zeros of $G(s)F(s)$ to the right of the point is odd.

For this example, we can easily find the only locus without any computation, using this rule. For verification of this rule, the angle at $s = -0.5$ is $\angle(s+1) = \angle(-0.5 + 1) = 0$ and root locus does not exist at that point. The angle at $s = -1.5$ is $\angle(s + 1) = \angle(-1.5 + 1) = -180°$ and root locus exists at that point.

Example 7.1.2 Figure 7.2a shows the root locus of $\frac{K}{(s+1)(s+2)}$. The location of the poles of the loop transfer function is indicated by crosses. One of the two root loci is shown by a dashed line for distinction. The arrows indicate the direction of increasing gain. According to Rule 1, the locus on the real axis has to be between -1 and -2. With increasing K, the constant term in the characteristic equation, which is quadratic for this example, also increases. From the formula for the solution of the quadratic equation, we know that the roots become complex-valued at some point. Then, with the real part of the roots fixed, the imaginary parts keep increasing and become a pair of vertical asymptotes. The locus should be checked at some test points for its correctness. For example, at $\sigma = \pm 0.5$, the phase is 0 and the points are not on the locus.

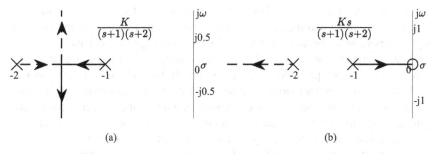

(a)   (b)

Fig. 7.2 (a) Root locus of $\frac{K}{(s+1)(s+2)}$; (b) root locus of $\frac{Ks}{(s+1)(s+2)}$

Rule 2 Asymptotes of the root loci The angles of the asymptotes, with $n \neq m$, are given by

$$\frac{180°(2k+1)}{|n-m|}, \quad n \neq m, \quad k = 0, 1, 2, \ldots, |n-m| - 1$$

The distinct angles are $\pm 90°$ in Fig. 7.2a. The **number of distinct asymptotes** is $n - m$ with $n \neq m$. If the number of asymptotes is odd, then one of them is on the negative real axis. For this example, for large values of s,

$$\frac{K}{(s+1)(s+2)} \approx \frac{K}{s^2} \to 0$$

There are two zeros at ∞.

Rule 3 Intersection of the asymptotes The intersection point is given by

$$\sigma_i = \frac{\Sigma \text{ real parts of poles of } G(s)F(s) - \Sigma \text{ real parts of zeros of } G(s)F(s)}{n - m}$$

The intersection, for this example, is at $((-1) + (-2))/2 = -1.5$.

The **breakaway point** is the point on the real axis segment between two poles, where they meet and become complex. At that point $s = \sigma$, the gain is maximum. The gain is given by

$$K = -\frac{1}{L(\sigma)} = -\frac{D(\sigma)}{N(\sigma)}$$

Therefore, the condition for a maximum point is

$$\frac{dK}{d\sigma} = \frac{d}{d\sigma}\left(\frac{D(\sigma)}{N(\sigma)}\right) = 0 \quad \text{or} \quad N(\sigma)D'(\sigma) - N'(\sigma)D(\sigma) = 0$$

For this example, the condition becomes $2s + 3 = 0$ and $s = -1.5$. In this case, the breakaway point and the point of intersection of the asymptotes are the same. Asymptotes show the behavior of the root loci for very high frequencies, as $\omega \to \infty$. A root locus may lie on one side of the corresponding asymptote or cross it.

Although software packages, such as MATLAB, has to be used to construct the root loci and other computations for practical systems, it is essential to do the plots and computations manually for better understanding of the concepts for simpler systems. One way to construct the root loci manually is to use the given set of rules. One brute force way of constructing the locus is to plot the roots of the characteristic equation, as the gain K varies from 0 to a desired value. Each root has a locus. Another brute force way of constructing the locus is to select an area around the origin in the s-plane, enclosing all the finite poles and zeros of $G(s)F(s)$. Now, select $N \times N$ discrete points in the area and apply the angle condition. All the points on the loci will have $180°$ phase, when the angles are expressed in their principle values in the range, $-\pi < \theta \leq \pi$. The number of points and the area can be changed to suit the particular $G(s)F(s)$. For this example, the 10×10 points in the loci area given in the following matrix clearly shows the root loci. The phase of the points those are not on the loci have been reduced to zero. As the loci is symmetric about the real axis, only the positive half of the phase values are shown.

$$
\begin{bmatrix}
0.9 & 0 & 0 & 0 & 0 & 0 & 180 & 0 & 0 & 0 & 0 \\
0.8 & 0 & 0 & 0 & 0 & 0 & 180 & 0 & 0 & 0 & 0 \\
0.7 & 0 & 0 & 0 & 0 & 0 & 180 & 0 & 0 & 0 & 0 \\
0.6 & 0 & 0 & 0 & 0 & 0 & 180 & 0 & 0 & 0 & 0 \\
0.5 & 0 & 0 & 0 & 0 & 0 & 180 & 0 & 0 & 0 & 0 \\
0.4 & 0 & 0 & 0 & 0 & 0 & 180 & 0 & 0 & 0 & 0 \\
0.3 & 0 & 0 & 0 & 0 & 0 & 180 & 0 & 0 & 0 & 0 \\
0.2 & 0 & 0 & 0 & 0 & 0 & 180 & 0 & 0 & 0 & 0 \\
0.1 & 0 & 0 & 0 & 0 & 0 & 180 & 0 & 0 & 0 & 0 \\
0 & 180 & 180 & 180 & 180 & 180 & 180 & 180 & 180 & 180 & 180 \\
 & -2 & -1.9 & -1.8 & -1.7 & -1.6 & -1.5 & -1.4 & -1.3 & -1.2 & -1.1
\end{bmatrix}
$$

Example 7.1.3 Figure 7.2b shows the root locus of $\frac{Ks}{(s+1)(s+2)}$. The location of the poles of the loop transfer function are indicated by crosses and the zero is indicated by a circle. According to Rule 1, one locus on the real axis has to be between -1 and 0. That is starting at the pole -1 and terminating at the zero 0. Another one is the asymptote from the pole at -2 going toward $-\infty$.

Example 7.1.4 Figure 7.3a shows the root locus of $\frac{K(s+0.5)}{s(s+1)(s+2)}$. According to Rule 1, one locus on the real axis has to be between -0.5 and 0. That is, starting at the pole 0 and terminating at the zero -0.5. The second one, starting at the pole -1, moves toward left. The third one, starting at the pole -2, moves toward right. The intersection of the asymptotes, for this example, is at $((-1) + (-2) - (-0.5))/2 = -1.25$. The angles are $\pm 90°$. Breakaway point, which must lie on the loci, must also satisfy the condition

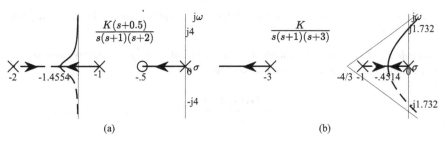

Fig. 7.3 (a) Root locus of $\frac{K(s+0.5)}{s(s+1)(s+2)}$; (b) root locus of $\frac{K}{s(s+1)(s+3)}$

$$\frac{d(G(s)F(s))}{ds} = 0$$

For this example,

$$\frac{d\left(\frac{(s+0.5)}{s(s+1)(s+2)}\right)}{ds} = 0$$

Differentiating and equating the derivative to zero, we get

$$2s^3 + 4.5s^2 + 3s + 1 = 0$$

The three roots of this equation are

$$\{-1.4554, -0.3973 + j0.4309, -0.3973 - j0.4309\}$$

The breakaway point, for this example, is expected between the points -2 and -1. Therefore, the breakaway point is -1.4554. From the breakaway point, the two branches entering the complex plane leave the real axis at angles of the asymptotes, $\pm 90°$. The **absolute value of** K at any point on the loci $s = s_0$ is given by

$$|K| = \frac{1}{|G(s_0)F(s_0)|}$$

Alternatively,

$$K = \frac{\text{product of lengths between point } s_0 \text{ to the poles}}{\text{product of lengths between point } s_0 \text{ to the zeros}}$$

For example,

$$|K| = \frac{1}{|G(s_0)|} = \left.\frac{|s_0 + 2| \times |s_0 + 1| \times |s_0|}{|s_0 + 0.5|}\right|_{s_0 = -0.499} = 375.2485$$

The point $s_0 = -0.499$ is on the root locus. As it is quite close to the zero, the gain is fairly high.

Example 7.1.5 Figure 7.3b shows the root locus of $\frac{1}{s(s+1)(s+3)}$. Breakaway point, which must lie on the loci, must also satisfy

$$\frac{d(G(s)F(s))}{ds} = 0$$

For this example,

$$\frac{d\left(\frac{1}{s(s+1)(s+3)}\right)}{ds} = 0$$

Differentiating and equating the derivative to zero, we get

$$3s^2 + 8s - 3 = 0$$

The two roots of this equation are

$$\{-2.2153, \qquad -0.4514\}$$

The breakaway point, for this example, is expected between the points 0 and -1. Therefore, the breakaway point is -0.4514. From the breakaway point, the two branches entering the complex plane leave the real axis at angles of $\pm 90°$. The angles of the three asymptote are $60°, \; 180°, 300°$. The characteristic equation is

$$s^3 + 4s^2 + 3s + K = 0$$

Let us find the **intersection of the root loci with the frequency axis**. On the frequency axis, $s = j\omega$. Replacing s by $j\omega$, we get

$$(j\omega)^3 + 4(j\omega)^2 + 3(j\omega) + K = 0 \text{ or } (K - 4\omega^2) + j(3\omega - \omega^3) = 0 + j0$$

Separating the real and imaginary parts of the equation, we get

$$K - 4\omega^2 = 0 \quad \text{and} \quad 3\omega - \omega^3 = \omega(3 - \omega^2) = 0$$

The frequencies are $\omega = 0$ and $\omega = \pm\sqrt{3}$, as shown in Fig. 7.3b. With $\omega = \sqrt{3}$, we get $K = 4\omega^2 = 12$.

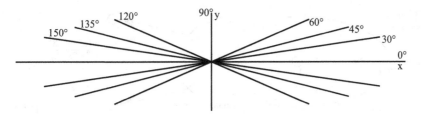

Fig. 7.4 Lines through the origin having various angles of inclination

Angle of a Line in the Complex Plane

Let us recollect the definition of the angle of a line in the complex plane. The angle of inclination of a nonhorizontal line is the smallest positive angle $\theta < \pi$ measured counterclockwise from the positive x-axis to the line. The angle for a horizontal line is $\theta = 0$ and $\theta = \frac{\pi}{2}$ for a vertical line. Angle θ for a line passing through two points with coordinates (x_1, y_1) and (x_2, y_2) is defined as

$$\tan(\theta) = \frac{y_2 - y_1}{x_2 - x_1} = \frac{y_1 - y_2}{x_1 - x_2}$$

Figure 7.4 shows lines through the origin having various angles of inclination. This figure is useful to get a rough estimation of angles of lines in determining angle of arrivals and departures of the loci.

Breakaway and Break-in Points

If the root locus branch lies on the real axis between two poles, then the locus must breakaway from real axis at some point between the two poles. Similarly, if the root locus branch lies on the real axis between two zeros, then the locus must make an entry into the real axis at some point between the two zeros.

Example 7.1.6 Figure 7.5a shows the root locus of $\frac{K(s+1)}{s^2+s+3}$. One locus has to start from a complex pole and terminate at $-\infty$. The other locus has to start from the other complex pole and terminate at the zero -1. **Break-in point**, which must lie on the loci, must also satisfy

$$\frac{d(G(s)F(s))}{ds} = 0$$

For this example,

$$\frac{d\left(\frac{(s+1)}{s^2+s+3}\right)}{ds} = 0$$

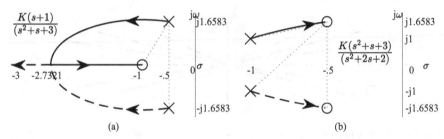

Fig. 7.5 (a) Root locus of $\frac{K(s+1)}{(s^2+s+3)}$; (b) root locus of $\frac{K(s^2+s+3)}{s^2+2s+2}$

Differentiating and equating the derivative to zero, we get

$$s^2 + 2s - 2 = 0$$

The two roots of this equation are

$$\{-2.7321, 0.7321\}$$

The break-in point, for this example, is expected to the left of the point -1. Therefore, the break-in point is -2.7321.

Angle of Departure of the Locus from a Complex Pole

The angle of departure (the direction in which the locus leaves a complex pole) is the angle of the tangent to the locus near the point. Let there be m poles $\{p_0, p_1, \ldots, p_{m-1}\}$ and n zeros $\{z_0, z_1, \ldots, z_{n-1}\}$ of $G(s)F(s)$. Let us find the angle of departure of the root locus from a complex pole at p_0.

1. Draw $(m + n - 1)$ lines between pole p_0 and the rest of the poles and zeros.
2. Let the corresponding angles of inclination from the lines to the other poles be $\{\theta_1, \theta_2, \ldots, \theta_{m-1}\}$.
3. Let the corresponding angles of inclination from the lines to the zeros be $\{\phi_0, \phi_1, \ldots, \phi_{n-1}\}$.
4. If the point $s = s_0$ is on the locus and very close to the complex pole p_0, then the angle of departure can be approximated to

$$180° + (\phi_0 + \phi_1 + \cdots + \phi_{n-1}) - (\theta_1 + \theta_2 + \cdots + \theta_{m-1})$$

As the pole is on the locus, the sum of the angle contributions of poles and zeros must be an integer multiple of $\pm 180°$.

For this example, the zero is at -1. The complex poles are at $-0.5 \pm j1.6583$. Let us find the angle of departure at the pole, $-0.5 + j1.6583$. As the line joining the complex-conjugate poles is a vertical line, $\theta_1 = 90°$. The angle of the line joining $-0.5 + j1.6583$ and the zero at $-1 + j0$ is

$$\frac{1.6583}{-0.5 - (-1)} = 3.3166 = \tan(\phi_0) \quad \text{or} \quad \phi_0 = 73.2212°$$

The angle of departure at pole $-0.5 + j1.6583$ is

$$180° + 73.2212° - 90° = 163.2212°$$

Angle of Arrival of the Locus at a Complex Zero
We follow the same procedure for finding the angle of departure, except that the angle of arrival (the direction in which the locus approaches a complex zero) is approximated to

$$180° - (\phi_1 + \phi_2 + \cdots + \phi_{n-1}) + (\theta_0 + \theta_1 + \cdots + \theta_{m-1})$$

Example 7.1.7 Figure 7.5b shows the root locus of $\frac{K(s^2+s+3)}{s^2+2s+2}$. The two poles are $(-1+j1)$ and $(-1-j1)$. The two zeros are $(-0.5+j1.6583)$ and $(-0.5-j1.6583)$. Let us find the angle of arrival at the zero, $-0.5 + j1.6583$. As the line joining the complex-conjugate zeros is a vertical line, $\phi_1 = 90°$. The angle of the line joining $-0.5 + j1.6583$ and the pole at $-1 + j1$ is

$$\frac{0.6583}{-0.5 - (-1)} = 1.3166 = \tan(\theta_0) \quad \text{or} \quad \theta_0 = 52.7822°$$

The angle of the line joining $-0.5 + j1.6583$ and the pole at $-1 - j1$ is

$$\frac{2.6583}{-0.5 - (-1)} = 5.3166 = \tan(\theta_1) \quad \text{or} \quad \theta_0 = 79.3477°$$

The angle of arrival at the zero $-0.5 + j1.6583$ is

$$180° + 52.7822° + 79.3477° - 90° = 222.1299°$$

Example 7.1.8 Figure 7.6a shows the root locus of $\frac{K(s+2)(s+3)}{s(s+1)}$. Two finite poles terminate at two finite zeros. On the real axis, two loci lie between the poles at 0 and at -1, and between the zeros at -2 and at -3. As gain increases, they break away at some point between 0 and -1. The loci break in at some point between -2 and -3. One locus starts at 0 and ends up at -2. Breakaway point, which must lie on the loci, must also satisfy

$$\frac{d(G(s)F(s))}{ds} = 0$$

For this example,

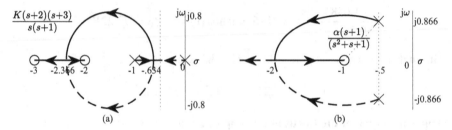

Fig. 7.6 (a) Root locus of $\frac{K(s+2)(s+3)}{s(s+1)}$; (b) root locus of $\frac{\alpha(s+1)}{(s^2+s+1)}$

$$\frac{d\left(\frac{(s^2+5s+6)}{(s^2+s)}\right)}{ds} = 0$$

Differentiating and equating the derivative to zero, we get

$$2s^2 + 6s + 3 = 0$$

The two roots of this equation are

$$\{-2.3660, -0.6340\}$$

The breakaway point, for this example, is expected between the points 0 and −1. Therefore, the breakaway point is −0.6340. The break-in point, for this example, is expected between the points −2 and −3. Therefore, the break-in point is −2.3660. With two poles and two zeros coming together, the angles at the breakaway and break-in points occur are the same and they are

$$\frac{\pm 180°}{2} = \pm 90°$$

Example 7.1.9 Figure 7.6b shows the root locus of $\frac{\alpha(s+1)}{s^2+s+1}$. If the variable parameter K does not appear as a multiplicative factor of $G(s)F(s)$, then it may be possible to rewrite the function to make K as a multiplicative factor of $G(s)F(s)$. Further, the variable parameter can be other than gain K also. Let

$$G(s)F(s) = \frac{1}{(s+1)(s+\alpha)}$$

Now, we want the locus as a function of the pole α. The characteristic equation is

$$s^2 + s + 1 + \alpha(s+1) = 0$$

Dividing both sides by the sum of the terms that do not contain α, we get

$$1 + \frac{\alpha(s+1)}{s^2 + s + 1} = 0$$

The open-loop zero at $s = -1$ resulted in modifying the characteristic equation to make α as a multiplicative factor.

Break-in point, which must lie on the loci, must also satisfy

$$\frac{d(G(s)F(s))}{ds} = 0$$

For this example,

$$\frac{d\left(\frac{(s+1)}{(s^2+s+1)}\right)}{ds} = 0$$

Differentiating and equating the derivative to zero, we get

$$s(s+2) = 0$$

The two roots of this equation are

$$\{0, -2\}$$

The break-in point, for this example, is expected to the left of the point $s = -1$. Therefore, the break-in point is at $s = -2$.

Angle of departure of the locus at the complex pole $s = -0.5 + j0.866$.

For this example, the zero is at $s = -1$. The complex poles are at $-0.5 \pm j0.866$. As the line joining the complex-conjugate poles is a vertical line, $\theta_1 = 90°$. The angle of the line joining $-0.5 + j0.866$ and the zero at $s = -1 + j0$ is

$$\frac{0.866}{-0.5 - (-1)} = 1.732 = \tan(\phi_0) \quad \text{or} \quad \phi_0 = 60°$$

The angle of departure at pole $-0.5 + j0.866$ is

$$180° + 60° - 90° = 150°$$

7.1.2 Nonminimum-Phase Systems

All the open-loop poles and zeros of a **minimum-phase** system lie in the left-half of the s-plane. Further, if both the system and its inverse are causal and stable, the system is of minimum-phase type. For example,

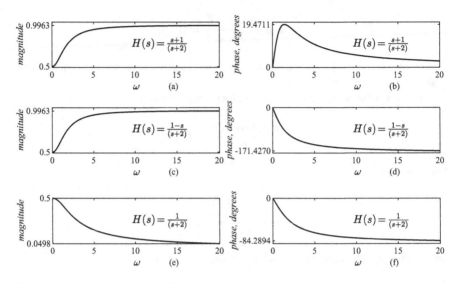

Fig. 7.7 Frequency responses of different types of systems

$$\frac{s+1}{s+2} \quad \text{and} \quad \frac{s+2}{s+1}$$

are causal and stable. At least one open-loop pole or zero of a **nonminimum-phase** system lies in the right-half of the s-plane. Let the transfer function $G(s)$ of a nonminimum-phase system be both causal and stable. Then, its inverse $1/G(s)$ is causal and unstable. Consider a transfer function and its inverse

$$\frac{1-s}{s+2} \quad \text{and} \quad \frac{s+2}{1-s}$$

The first one is causal and stable, while the second one is causal and unstable. Figure 7.7 shows the frequency responses of different types of systems. Figure 7.7a and b show, respectively, the magnitude and phase of the frequency response of a minimum-phase system. The variation in the phase is lowest. Figure 7.7c and d show, respectively, the magnitude and phase of the frequency response of a nonminimum-phase system. The variation in the phase is largest. The magnitude response is the same. Figure 7.7e and f show, respectively, the magnitude and phase of the frequency response of a system, which does not fit into the two types defined. The variation in the phase is between those of the two types defined.

A system associated with delay units are of nonminimum-phase type. For a given magnitude response, there is only one minimum-phase system and it is not case for nonminimum-phase systems. There is an undershoot in the step response of nonminimum-phase systems. When we open the hot water supply in a water heating system, the water is cold initially and it takes some time to get hot water. It is a nonminimum-phase system. Consider the open-loop transfer function of a

Fig. 7.8 Root locus of $\frac{K(1-s)}{s(s+2)}$

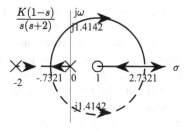

nonminimum-phase system

$$G(s) = \frac{K(1-s)}{s(s+2)} = -\frac{K(s-1)}{s(s+2)}$$

with unity feedback. Let us plot the root locus of this system. On a given section of the real axis, the root locus exists only if the sum of the number of all the poles and zeros, located on the real axis, of $G(s)$ to the right of the section is even, as shown in Fig. 7.8. The angle condition is that

$$\angle\left(\frac{K(s-1)}{s(s+2)}\right) = 0$$

There is an asymptote starting from the pole at $s = -2$. By differentiating $G(s)$ and equating the derivative to 0, we get the break-in and breakaway points as 2.7321 and -0.7321, respectively. Let us find the point where the root locus crosses the imaginary axis. The characteristic equation of the system is

$$s^2 + 2s - Ks + K = 0$$

Replacing s by $j\omega$, we get

$$-\omega^2 + 2j\omega - Kj\omega + K = 0$$

Equating real and imaginary parts of the equation to 0, we get $\omega = \sqrt{2}$ and $K = 2$.

7.2 Control System Design by Root Locus Method

Given $G(s)F(s)$, the design process has to ensure that the corresponding closed-loop system is stable with sufficient stability margin and has specified response. Another factor to consider is the order of system with capability to meet the specifications. The specifications must be realistic with the given system. For example, an electric motor will stall if it is loaded more than its load capability. A first-order system cannot give oscillatory response, since its only pole is real for practical systems. Another example is that the attenuation a first-order Butterworth filter can provide is limited to 20 dB/decade. For larger attenuations, higher-order filters are required. Correction means treatment for a specific defect. If we are overweight, the correction is to change our eating and exercise habits suitably. Similarly, if the system response is not the desired one, then the system has to be modified suitably. There are two types of corrections commonly used to make the response of a control system acceptable. One type is to change the gain K. As the gain K is increased from zero, the system, with a pair of real poles located in the left-half of the s-plane, remains overdamped within some range (the locus moves along the real axis and the roots are real). For further increase of K, the roots become complex-conjugates and the system becomes underdamped and gives oscillatory response with a gradual increase in the frequency of oscillation and overshoot.) The design of a control system is to modify its transfer function suitably so that the performance of the system is as desired. It is convenient to change the transfer function using tools such as root locus, Bode, and Nyquist plots.

Another type of correction is to add poles and zeros to the given $G(s)F(s)$. This correction modifies the root locus. While we can modify the root locus in any desired way, it has to be ensured that the angle condition is always maintained. A unit, called compensator, is added in the forward path, called series compensation, to make the loop transfer function

$$G_c(s)G(s)F(s)$$

where $G_c(s)$ is the transfer function of the compensator. By changing the gain and adding the compensator unit, the root locus can be changed suitably. If a single stage of the compensator is inadequate, more stages can be suitably added. Further, the design procedure involves some number of iterations and it is not unique. The number of iterations increases with less number of constraints on the design. Although the compensator can also be placed in the feedback path, we concentrate on series compensation only. The design procedure is better presented through examples. *A compensator, composed of poles and zeros, changes the response of the compensated system appropriately, so that the response is the desired one with sufficient stability margin.*

7.2.1   Proportional Compensator

Figure 7.9 shows a feedback system with a proportional compensator. In this type of compensator, the control signal at the output of the compensator is K_p times of its input. This constant and the undamped natural frequency are controllable, but not the damping ratio of the system. The result is that adequate steady-state error can be obtained by varying K_p, but with poor transient response.

Fig. 7.9 Feedback system
with a proportional
compensator

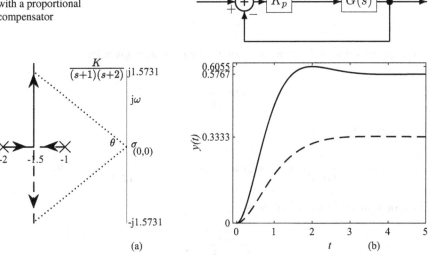

Fig. 7.10 (a) Root locus of $\frac{K}{(s+1)(s+2)}$; (b) unit-step response of $H(s) = \frac{1}{s^2+3s+3}$ and $H(s) = \frac{2.7246}{s^2+3s+4.7246}$

Example 7.2.1 Consider the loop transfer function

$$\frac{K}{(s+1)(s+2)}$$

of a unity feedback system. The design specification is that the overshoot M_p is to be limited to 5%. Use the proportional compensator.

Solution Figure 7.10a shows the root locus of $\frac{K}{(s+1)(s+2)}$. From the definition of M_p, we get

$$\zeta = \frac{-\log_e(M_p)}{\sqrt{\pi^2 + \log_e^2(M_p)}} \geq 0.6901$$

The damping ratio ζ of a pair of complex-conjugate poles is related to the angle θ between the negative real axis and the line joining the origin and the location of the complex poles. That is,

$$\zeta = \cos(\theta)$$

The line intercepts the root locus at

$$y = \omega_d = x \tan(\theta) = x \tan(\cos^{-1}(\zeta))$$

The frequencies of interception, for this example, are

$$-1.5 \pm j1.5731$$

The absolute value of K at any point on the loci $s = s_0$ is given by

$$|K| = \frac{1}{|G(s_0)|}$$

For this example,

$$|K| = \frac{1}{|G(s_0)|} = |s_0 + 2| \times |s_0 + 1| \Big|_{s_0 = -1.5 + j1.5731} = 2.7246$$

The loop transfer function becomes

$$\frac{2.7246}{(s^2 + 3s + 2)}$$

The corresponding closed-loop transfer function is

$$H(s) = \frac{2.7246}{s^2 + 3s + 4.7246}$$

The lines of constant ζ are shown by dotted lines, in Fig. 7.10a. They are radial lines passing through the origin with angle θ and $-\theta$ from the negative real axis. For a given ζ, the points of intersection of them with the root locus give the closed-loop pole locations of the designed system. The gain constant K_p can be computed at these frequencies. The unit-step responses before (dashed line) and after compensation are shown in Fig. 7.10b. The overshoot specification has been met. The steady-state response is 0.5767, as can be seen from the figure. Analytically, the steady-state output value, for unit-step input signal, is obtained by applying the final value theorem to the output expression $Y(s)$ to get the final output value.

$$\lim_{s \to 0} \frac{2.7246s}{s(s^2 + 3s + 4.7246)} = 0.5767$$

We can increase the steady-state value by increasing the gain further. But, it will degrade the transient response by decreasing the damping ratio. That is, the system becomes more underdamped. With gain constant alone, we can set the steady-state error or overshoot, but not both simultaneously.

Fig. 7.11 Feedback system with a proportional-integral compensator

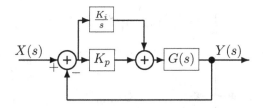

7.2.2 Proportional-Integral Compensator

The compensator called proportional-integral (PI) compensator is shown in Fig. 7.11. That is, we have to add a zero and a pole at 0. The pole is located to the right of the zero. This compensator is essentially a lowpass filter. The compensator is of the form

$$K\frac{(s+a)}{s} = K + \frac{Ka}{s} = K_p + \frac{K_i}{s}$$

The control input, now, has two components: (a) one is proportional to the error signal and (b) another is proportional to the integral of the error signal. The pole at the origin reduces the steady-state error. The nearby zero almost cancels the effect of the pole, leaving the transient response nearly unchanged. The zero is placed very close to the pole at the origin so that the gain and angular contribution of this pole-zero pair to the root locus is almost zero. This compensator does not affect the transient response much but drastically improves the steady-state error due to the increase in system type by one.

Example 7.2.2 Consider the loop transfer function

$$\frac{1}{(s+1)(s+2)}$$

of a unity feedback system. The design specification is that the overshoot of the unit-step response is to be limited to 5%. Further, the steady-state error is to be reduced. Use the PI compensator.

Solution To control both the steady-state error and overshoot, the root locus of the given loop transfer function has to be modified. Let the zero be $(s + 0.1)$. The magnitude and angle of the compensator at the desired frequency are 0.9682 and $-2.0111°$, respectively. Therefore, its impact on the transient response is negligible since neither the change in gain nor in phase is significant and ignored in the design. One can experiment with different zero locations. Including that of the plant, we get loop transfer function of the compensated system as

$$K\frac{(s+0.1)}{s}\frac{1}{(s+1)(s+2)} = K\frac{(s+0.1)}{s(s+1)(s+2)}$$

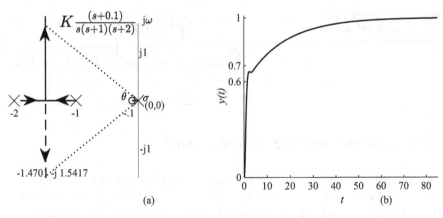

Fig. 7.12 (a) Root locus of $K\frac{(s+0.1)}{s(s+1)(s+2)}$; (b) unit-step response of $H(s) = \frac{2.7139s+0.2714}{s^3+3s^2+4.7139s+0.2714}$

Figure 7.12a shows the root locus of this loop transfer function by a thick line. Breakaway point, which must lie on the loci, must also satisfy

$$\frac{d(G(s)F(s))}{ds} = 0$$

For this example,

$$\frac{d\left(\frac{(s+0.1)}{s(s+1)(s+2)}\right)}{ds} = 0$$

Differentiating and equating the derivative to zero, we get

$$2s^3 + 3.3s^2 + 0.6s + 0.2 = 0$$

The three roots of this equation are

$$\{-1.4940, \qquad -0.0780 \pm j0.2467\}$$

The breakaway point, for this example, is expected between the points -2 and -1. Therefore, the breakaway point is -1.4940.

From the definition of M_p, we get

$$\zeta = \frac{-\log_e(M_p)}{\left(\sqrt{\pi^2 + \log_e^2(M_p)}\right)} \geq 0.6901$$

The frequencies of interception found graphically from the root locus, for this example, are

$$-1.4701 \pm j1.5417$$

The absolute value of K at any point on the loci $s = s_0$ is given by

$$|K| = \frac{1}{|G(s_0)|}$$

For this example,

$$|K| = \frac{1}{|G(s_0)|} = \left. \frac{|s_0 + 2| \times |s_0 + 1| \times |s_0 + 0|}{|s_0 + 0.1|} \right|_{s_0 = -1.4701 + j1.5417} = 2.7139$$

The loop transfer function is

$$\frac{2.7139s + 0.2714}{s(s^2 + 3s + 2)}$$

The corresponding closed-loop transfer function becomes

$$H(s) = \frac{2.7139s + 0.2714}{s^3 + 3s^2 + 4.7139s + 0.2714}$$

The step response is shown in Fig. 7.12b.

7.2.3 Lag Compensator

In general, adding a pole in the left-half of the s-plane moves the root locus to the right. The consequences are:

1. The relative stability of the system is reduced.
2. The response becomes slower.
3. The steady-state error is reduced.

In general, adding a zero in the left-half of the s-plane moves the root locus to the left. The consequences are:

1. The relative stability of the system is increased.
2. The response becomes faster.
3. The steady-state error is increased.

Lag compensator increases the gain at low frequencies and, hence, reduces the steady-state error. The steady-state error constants are multiplied by the factor z_c/p_c due to the compensator and, hence, the steady-state error is reduced. The phase of

Fig. 7.13 Feedback system
with a lag compensator

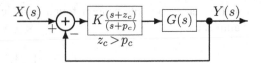

the compensator is always negative, justifying its name. The difference between the PI and lag compensators is that the steady-state error is eliminated by the PI compensator, while it is reduced by the lag compensator. The lag compensator reduces the gain of the system at high frequencies and, thereby, attenuates the high frequency noise better. This compensator shifts the root locus to the right slightly.

Example 7.2.3 Consider the loop transfer function

$$\frac{K}{(s+1)(s+2)}$$

of a unity feedback system. The design specification is that the overshoot is to be limited to 5%. Further, let the zero of the compensator z_c be 10 times that of the pole p_c. Use the lag compensator.

Solution The compensator does not change the type of the system. The essential difference between the PI compensator and this is that the pole is located near the origin, but not exactly at the origin. Figure 7.13 shows the block diagram of the lag compensator. That is, we have to add a zero and a pole close to the origin. The pole is located to the right of the zero. This compensator is also basically a lowpass filter. The PI compensator requires active circuits, such as the operational amplifier circuit that provides a very large gain to simulate an ideal integrator response $1/s$. However, the lag compensator can be realized with passive circuit components, such as resistors and capacitors. The compensator is of the form

$$K\frac{(s+z_c)}{(s+p_c)}, \qquad p_c < z_c$$

As $s \to 0$,

$$\lim_{s \to 0} K\frac{(s+z_c)}{(s+p_c)} = K\frac{z_c}{p_c}$$

The position error constant is increased by the factor Kz_c/p_c. Let the pole and the zero be $(s+0.01)$ and $(s+0.1)$, respectively. As already noted, the design involves trial-and-error. Then, the transfer function of the compensator is

$$K\frac{(s+0.1)}{(s+0.01)}$$

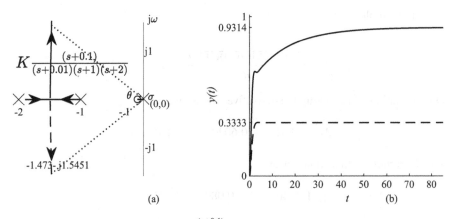

Fig. 7.14 (a) Root locus of $K\frac{(s+0.1)}{(s+0.01)(s+1)(s+2)}$; (b) unit-step response of $H(s)$ = $\frac{2.7156s+0.2716}{s^3+3.01s^2+4.7456s+0.2916}$

With the gain 2.7156, as found later, the error constant is increased by the factor 27.156. The position error constant of the given system is 0.5. Therefore, the position error constant of the compensated system becomes $27.156\times0.5 = 13.5780$. Then, the steady-state error is

$$e_{ss} = \frac{1}{1+13.5780} = 0.0686$$

The steady-state response of the compensated system is $1 - 0.0686 = 0.9314$, as shown in Fig. 7.14b. The steady-state error to the unit-step response is considerably decreased. To avoid the difference between the root loci of that of the compensated and uncompensated systems, the angle contribution of the lag compensator should be limited to $-5°$. Therefore, the pole and zero of the lag compensator should be placed relatively close together and near the origin. Then, the transient response will not be affected much. The gain and angle contribution of the lag compensator, for the present example, is 0.9714 and $-1.8118°$, respectively, and, therefore, ignored in the design. Including that of the plant, we get loop transfer function of the compensated system as

$$K\frac{(s+0.1)}{(s+0.01)}\frac{1}{(s+1)(s+2)} = K\frac{(s+0.1)}{(s+0.01)(s+1)(s+2)}$$

Figure 7.14a shows the root locus of this loop transfer function. Breakaway point, which must lie on the loci, must also satisfy

$$\frac{d(G(s)F(s))}{ds} = 0$$

For this example,

$$\frac{d\left(\frac{(s+0.1)}{(s+0.01)(s+1)(s+2)}\right)}{ds} = 0$$

Differentiating and equating the derivative to zero, we get

$$2s^3 + 3.31s^2 + 0.6020s + 0.1830 = 0$$

The three roots of this equation are

$$\{-1.4946, \qquad -0.0802 \pm j0.2341\}$$

The breakaway point, for this example, is expected between the points -2 and -1. Therefore, the breakaway point is -1.4946. From the definition of M_p, we get

$$\zeta = \frac{-\log_e(M_p)}{\sqrt{\pi^2 + \log_e^2(M_p)}} \geq 0.6901$$

The frequencies of interception of the constant ζ line with the locus, for this example, are

$$-1.473 \pm j1.5451$$

These values are found graphically, as the root locus is bent toward right. The absolute value of K at any point on the loci $s = s_0$ is given by

$$|K| = \frac{1}{|G(s_0)|}$$

For this example,

$$|K| = \frac{1}{|G(s_0)|} = \frac{|s_0 + 2| \times |s_0 + 1| \times |s_0 + 0.01|}{|s_0 + 0.1|}\Big|_{s_0 = -1.473 + j1.5451} = 2.7156$$

The loop transfer function is

$$\frac{2.7156s + 0.2716}{(s + 0.01)(s^2 + 3s + 2)}$$

The corresponding closed-loop transfer function becomes

$$H(s) = \frac{2.7156s + 0.2716}{s^3 + 3.01s^2 + 4.7456s + 0.2916}$$

Fig. 7.15 Feedback system with a proportional-derivative compensator

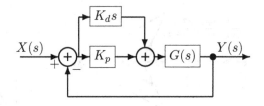

Figure 7.14b shows the step response before (dashed line) and after compensation.

7.2.4 Proportional-Derivative Compensator

Example 7.2.4 Consider the loop transfer function

$$\frac{K}{(s+1)(s+2)(s+4)}$$

of a unity feedback system. The design specification is that the overshoot is to be limited to 5%. Further, the settling time t_s is to be about 2 s. Use the proportional-derivative compensator.

Solution This compensator does not change the type of the system. This compensator adds a zero to the open-loop transfer function of the plant. The transfer function of this compensator is

$$K(s+z_c) = K_p + K_d s$$

Figure 7.15 shows this compensator. This compensator is basically a highpass filter. The PD compensator requires active circuits, such as the operational amplifier circuit that provides a very large gain to simulate an ideal differentiator response s. Let us determine where the zero has to be located to meet the specifications. From the previous example, $\zeta = 0.6901$. With $t_s = 2$, $\sigma = 4/2 = 2$. $\omega_n = \sigma/\zeta = 2/0.6901 = 2.8981$. The characteristic equation is

$$s^2 + 2\sigma s + \omega_n^2 = s^2 + 4s + 8.3992$$

The two roots of this equation are $-2 \pm j2.0974$. The transfer function of the compensated system is

$$\frac{K(s+z_c)}{(s+1)(s+2)(s+4)}$$

Let us find the condition that the two points $-2 \pm j2.0974$ are located on the root locus. The phase of the root locus must be 180° at $s_o = -2 + j2.0974$. Then,

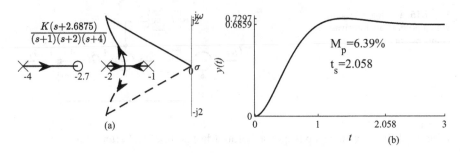

Fig. 7.16 (a) Root locus of $K\frac{6.5(s+2.6875)}{(s^3+7s^2+14s+8)}$; (b) unit-step response of $H(s)$ = $\frac{6.5(s+2.6875)}{(s^3+7s^2+20.5s+25.47)}$

$$\angle\left(\frac{K(s_o+z_c)}{(s_o+1)(s_o+2)(s_o+4)}\right) = 180°$$

The angle contribution of the all the poles at this frequency is $-251.8526°$. Therefore, the zero has to contribute $71.8526°$.

$$71.8526° = \tan^{-1}\left(\frac{2.0974}{z_c-2}\right) \quad \text{or} \quad \tan(71.8526°) = 3.051 = \left(\frac{2.0974}{z_c-2}\right)$$

Solving for z_c, we get the compensator as

$$(s+2.6875)$$

Including that of the plant, we get the loop transfer function of the compensated system as

$$K\frac{(s+2.6875)}{(s+1)(s+2)(s+4)}$$

Figure 7.16a shows the root locus of this loop transfer function. For this example, the point of intersection is -1.5587, obtained by differentiating the loop transfer function and equating the derivative to 0

The absolute value of K at any point on the loci $s = s_0$ is given by

$$|K| = \frac{1}{|G(s_0)|}$$

For this example,

$$|K| = \frac{1}{|G(s_0)|} = \frac{|s_0+2| \times |s_0+1| \times |s_0+4|}{|s_0+2.6875|}\Bigg|_{s_0=-2+j2.0974} = 6.5$$

Fig. 7.17 Feedback system
with a lead compensator

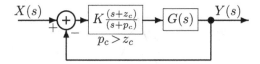

The loop transfer function is

$$\frac{6.5(s + 2.6875)}{(s^3 + 7s^2 + 14s + 8)}$$

The corresponding closed-loop transfer function becomes

$$\frac{6.5(s + 2.6875)}{(s^3 + 7s^2 + 20.5s + 25.47)}$$

The unit-step response is shown in Fig. 7.16b. The overshoot is little higher than 5%. As always, iterations are required to further refine the response. For this example, designing with overshoot 4% and setting time 1.8 s yield an overshoot 5.05% and setting time 1.94 s.

Higher-Order Systems

The design of control systems using the root locus is based on second-order transfer function. If only two dominant poles contribute significantly to the transient response of a higher-order system, then its response can be approximated by a second-order system. The closed-loop poles, which have the strongest influence on the transient response, are called the **dominant** closed-loop poles. Both the real part of the pole and the magnitude of its residue in the partial-fraction expansion of the transfer function evaluated at the pole determine the relative dominance.

In all other cases, the transfer function can be decomposed into first- and second-order systems using partial-fraction expansion and, then, each section can be separately analyzed.

7.2.5   Lead Compensator

The lead compensator, shown in Fig. 7.17, is the same as the lag compensator with $p_c > z_c$. It improves the transient response.

Example 7.2.5 Consider the loop transfer function

$$\frac{K}{s(s + 1)}$$

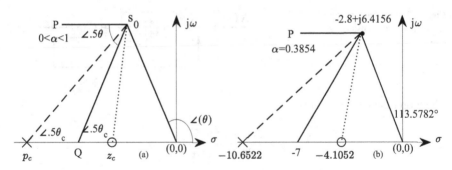

Fig. 7.18 Locating lead compensator pole and zero graphically

of a unity feedback system. The design specification is that the velocity error constant

$$K_v = \lim_{s \to 0} sG_c(s)L(s) \ge 20$$

The dimensionless damping ratio ζ is given as 0.4. Use the lead compensator to design the system.

Solution The transfer function of this compensator is

$$K\frac{(s + z_c)}{(s + p_c)}, \qquad p_c > z_c$$

This compensator is basically a highpass filter. The PD compensator requires active circuits, such as the operational amplifier circuit that provides a very large gain to simulate an ideal differentiator response s. The lead compensator can be implemented with passive components. Let us determine where the compensator zero and pole have to be located to meet the specifications. The desired poles have to be far from the loop transfer function poles.

A suggested procedure for locating the pole and zero is as follows. In Fig. 7.18, one of the desired pole location s_0 is shown and the design procedure is also illustrated. The compensator pole p_c and zero z_c are also indicated. Draw a line connecting the origin to the location of the pole. Let the angle subtended by this line be θ. Draw a horizontal line between s_0 and the point P. Bisect the angle between the two lines we have drawn and draw the bisector crossing the horizontal axis at the point Q. Let θ_c be the phase required from the compensator to make the angle of the loop transfer function an odd multiple of $180°$. Draw two lines $s_0 z_c$ and $s_0 p_c$ to make angles $0.5\theta_c$ from the bisector $s_0 Q$ on both of its sides. The intersection points of these lines with the negative real axis are the locations of the zero and pole of the compensator, as shown in the figure. Therefore, once the desired pole locations and the required angle from the compensator known, the locations of the compensator pole and zero can be easily determined graphically without any computation. The

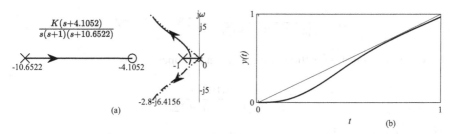

Fig. 7.19 (a) Root locus of $K \dfrac{(s+4.1052)}{s^3+11.6522s^2+10.6522s}$; (b) unit-ramp response of $H(s) =$ $\dfrac{72.24s+296.5596}{(s^3+11.6522s^2+82.8922s+296.5596)}$

locations can also be determined analytically, as shown in the specific example. Let us define $\alpha = Z_c/p_c$. The suggested procedure yields the largest value for α, resulting in a larger error constant. As Z_c is more closer to the origin than p_c, the value of α is constrained to be between 0 and 1, as shown in the figure.

The angular location of the desired complex-conjugate poles is determined by the damping ratio ζ. The distance of the poles from the origin is determined by the undamped natural frequency ω_n. The procedure is trial-and-error. Increasing the natural frequency ω_n pushes the desired poles further. After trying with some ω_n, $\omega_n = 7$ is chosen. Then, with $\zeta = 0.4$, we get the desired poles as $-2.8 \pm j6.4156$, as shown in Fig. 7.19a. Referring to Fig. 7.18b, for this example,

$$\theta = \angle(-2.8 + j6.4156) = 113.5782° \quad \text{and} \quad 0.5\theta = 56.7891°$$

The angle of the given system at the desired frequency is

$$\angle\left(\frac{1}{s(s+1)}\right)\Big|_{s=(-2.8+j6.4156)} = -219.2505°$$

To force the root locus pass through the closed-loop pole, the lead compensator must contribute $39.2505°$ to make total angle $-180°$. Let us recollect the point-slope form of the equation of a straight line. An equation for the line through the point (x_1, y_1), with slope m, is

$$y - y_1 = m(x - x_1)$$

Using this equation, we get

$$p_c = \frac{-6.4156}{\tan(39.2505°)} - 2.8 = -10.6522$$

Similarly,

$$z_c = \frac{-6.4156}{\tan((2 \times 39.2505)°)} - 2.8 = -4.1052$$

Now, the loop transfer function of the plant and compensator is

$$K \frac{(s + 4.1052)}{s(s + 1)(s + 10.6522s)}$$

The absolute value of K at any point on the loci $s = s_0$ is given by

$$|K| = \frac{1}{|G(s_0)|}$$

For this example,

$$|K| = \frac{1}{|G(s_0)|} = \left. \frac{|s_0 + 0| \times |s_0 + 1| \times |s_0 + 10.6522|}{|s_0 + 4.1052|} \right|_{s_0 = -2.8 + j6.4156} = 72.24$$

The loop transfer function is

$$72.24 \frac{(s + 4.1052)}{s^3 + 11.6522s^2 + 10.6522s}$$

The corresponding closed-loop transfer function becomes

$$H(s) = \frac{72.24s + 296.5596}{(s^3 + 11.6522s^2 + 82.8922s + 296.5596)}$$

The root locus is shown in Fig. 7.19a. The unit-ramp response is shown in Fig. 7.19b. The velocity error constant

$$K_v = \lim_{s \to 0} sG_c(s)L(s) = \frac{72.24(4.1052)}{10.6522} = 27.8402$$

Alternate Design

The number of possibilities to select the locations of the compensator pole and zero is unlimited. Further, if the angle to be contributed by the compensator is more than 65°, additional compensator sections may be added. If the previous design is not satisfactory in some respect, then place the zero on the real axis just below or slightly to the left of the desired poles. Let $z_c = -2.9$. The poles of the loop transfer function are 0 and -1. Applying the angle condition for the desired poles to be on the locus, we get

$$180 = \angle(s_o + 2.9) - \angle(s_o + 0) - \angle(s_o + 1) - \angle(s_o + p_c)$$

where $s_o = -2.8 + j6.4156$. Solving this equation for $\angle(s_o + p_c)$, we get it as $49.8565°$. Now,

$$p_c = -2.8 - \frac{6.4156}{\tan(49.8565°)} = -8.2108$$

The absolute value of K at any point on the loci $s = s_0$ is given by

$$|K| = \frac{1}{|G(s_0)|}$$

For this example,

$$|K| = \frac{1}{|G(s_0)|} = \left.\frac{|s_0 + 0| \times |s_0 + 1| \times |s_0 + 8.2108|}{|s_0 + 2.9|}\right|_{s_0 = -2.8 + j6.4156} = 61.0096$$

The loop transfer function is

$$61.0096\frac{(s + 2.9)}{s(s + 1)(s + 8.2108)}$$

The corresponding closed-loop transfer function becomes

$$H(s) = \frac{61.0096(s + 2.9)}{(s^3 + 9.211s^2 + 69.22s + 176.9)}$$

The root locus is shown in Fig. 7.20a. The closed-loop poles of the compensated system are

$$\{-3.6102, \quad -2.8 \pm j6.4156\}$$

The response of the designed system should be checked with standard signals to ensure that it meets the specifications. The unit-ramp response is shown in Fig. 7.20b. The velocity error constant

$$K_v = \lim_{s \to 0} sG_c(s)L(s) = \frac{61.0096(2.9)}{8.2108} = 21.5482$$

As always, iterations are required to further refine the response.

7.2.6 Proportional-Integral-Derivative Compensator

Lead and PD compensators improve the transient response. Lag and PI compensators improve the steady-state error. In the PID compensators, both the steady-state

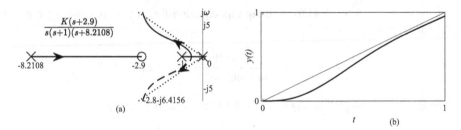

Fig. 7.20 (a) Root locus of $K\frac{(s+2.9)}{s(s+1)(s+8.2108)}$; **(b)** unit-ramp response of $H(s) = \frac{61.01(s+2.9)}{(s^3+9.211s^2+69.22s+176.9)}$

Fig. 7.21 Feedback system with a PID compensator

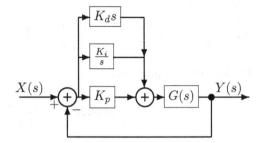

and transient responses are improved. The proportional-integral-derivative (PID) compensator, shown in Fig. 7.21, is the same as the PI compensator with another compensator component that is the derivative of the error signal. We can design for the transient response first and then design for the steady-state response or vice versa. In either case, the first design will be affected by the second design. As always, iterations are required to finalize the design. In the following example, let us design for the transient response first. The transfer function of the compensator is

$$G_c(s) = K_p + \frac{K_i}{s} + K_d s = \frac{K_d s^2 + K_p s + K_i}{s}$$

There are two zeros and a pole at the origin. One of the two zeros is determined through the PD design and the other through the PI design. The system type is increased due to the pole at the origin.

Example 7.2.6 Consider the loop transfer function

$$\frac{K}{(s+1)(s+2)(s+4)}$$

of a unity feedback system. The design specification is that the overshoot is to be limited to about 15% and settling time is to be about 2 s. Use the PID compensator to design the system.

Solution Let the settling time be about 1.5 s. This way, we can reduce the number of iterations.

$$\sigma = \frac{4}{1.5} = \frac{8}{3}, \quad \zeta = \frac{-\log(0.15)}{\sqrt{\pi^2 + \log^2(0.15)}} = 0.5169, \quad \omega_d = \frac{\sigma}{\zeta}\sqrt{1 - \zeta^2} = 4.4159$$

The desired poles are $-2.6667 \pm j4.4159$.

Let us design the PD compensator first. The transfer function of this compensator is

$$K(s + z_c)$$

Let us determine where the zero has to be located to meet the specifications. The phase angle of the given system at the desired frequency $-2.6667 + j4.4159$ is

$$(-\angle(s + 1) - \angle(s + 2) - \angle(s + 4)) = -282.4629°$$

Since this angle has to be an odd multiple of 180°, the lead compensator has to provide 102.4629°. However, the maximum phase lead that may be produced by a practical PD compensator is about 65°. Therefore, we need two stages of lead compensation to provide the necessary phase. One set of arbitrary choices for the location of the zeros are -5.5 and -7.0599. Including that of the plant, we get the loop transfer function of the PD compensated system as

$$K\frac{(s + 5.5)(s + 7.0599)}{(s + 1)(s + 2)(s + 4)}$$

The gain constant K, with $s_0 = -2.6667 + j4.4159$, is determined as

$$|K| = \frac{|s_0 + 1| \times |s_0 + 2| \times |s_0 + 4|}{|s_0 + 5.5| \times |s_0 + 7.0599|}\bigg|_{s_0 = -2.6667 + j4.4159} = 2.9752$$

The root locus is shown in Fig. 7.22a. The PD compensator design was verified to meet the transient response specifications. Otherwise, iterate.

Now, we have to add the PI compensator to the designed PD compensator. That is, we have to add a pole at 0 and a zero nearby. The pole is located to the right of the zero. The PI compensator is of the form

$$\frac{(s + a)}{s}$$

The major purpose of the PI compensator is make the steady-error zero leaving the transient response almost unchanged. The value of a is found by trying various values to the left of the pole at origin. After ensuring that the total response is acceptable, the compensator is found to be $(s + 0.3)/s$. The magnitude and phase of $(s + 0.3)/s$ at $-2.6667 + j4.4159$ are 0.9712 and $-2.9381°$, respectively, and small

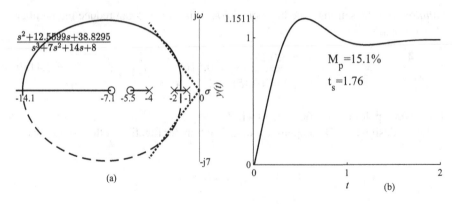

Fig. 7.22 (a) Root locus of PD compensated system; (b) unit-step response of $H(s)$

enough to be ignored. The gain has been increased by a factor of 1.1 to get 3.2727 to improve the final response. Including that of the plant, we get the loop transfer function of the PID compensated system as

$$K\frac{(s+5.5)(s+7.0599)(s+0.3)}{s(s+1)(s+2)(s+4)} = \frac{3.2727s^3 + 42.0865s^2 + 139.4083s + 38.1231}{s^4 + 7s^3 + 14s^2 + 8s}$$

The corresponding closed-loop transfer function becomes

$$H(s) = \frac{3.2727s^3 + 42.0865s^2 + 139.4083s + 38.1231}{s^4 + 10.27s^3 + 56.09s^2 + 147.4s + 38.12}$$

The unit-step response after compensation is shown in Fig. 7.22b with performance figures.

7.2.7   Lead-Lag Compensator

Lead compensator improves the transient response. Lag compensator improves the steady-state error. In the lead-lag compensator, both the steady-state and transient responses are improved. The lead-lag compensator, shown in Fig. 7.23, is a combination of the lead and lag compensators. The transfer function of this compensator is

$$G_c(s) = K\frac{(s + z_{lead})}{(s + p_{lead})}\frac{(s + z_{lag})}{(s + p_{lag})}$$

In the lead compensator, $p_{lead} > z_{lead}$. In the lag compensator, $p_{lag} \approx 0 < z_{lag}$. The PID compensator requires active circuits, such as the operational amplifier

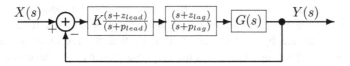

Fig. 7.23 Feedback system with a lead-lag compensator

circuit that provides a very large gain to simulate an ideal integrator response $1/s$. The lead-lag compensator can use passive components.

Example 7.2.7 Consider the loop transfer function

$$\frac{K}{(s+1)(s+2)(s+3)}$$

of a unity feedback system. The design specification is that the overshoot is to be limited to about 15%, the steady-state error is to be about 10% with respect to a constant reference and the settling time is to be about 3 s. Use the lead-lag compensator to design the system.

Solution The specification of the settling time is reduced to 2 s in order to reduce the number of iterations.

$$\sigma = \frac{4}{2} = 2, \quad \zeta = \frac{-\log(0.15)}{\sqrt{\pi^2 + \log^2(0.15)}} = 0.5169, \quad \omega_d = \frac{\sigma}{\zeta}\sqrt{1-\zeta^2} = 3.312$$

The desired poles are located at $-2 \pm j3.312$. Let us design the lead compensator first.

A lead compensator has a pole to filter out the high frequency noise. The effect of a pole located at higher frequencies does not affect the response of the compensator much. The transfer function of this compensator is

$$K\frac{(s + z_{lead})}{(s + p_{lead})}$$

Let us determine where the zero and pole have to be located to meet the specifications. The phase angle of the given system at the desired frequency $-2 + j3.312$ is

$$(-\angle(s + 1) - \angle(s + 2) - \angle(s + 3)) = -270°$$

Since this angle has to be an odd multiple of 180°, the lead compensator has to provide 90°. However, the maximum phase lead that may be produced by a practical lead compensator is about 65°. Therefore, we need two stages of lead compensation to provide the necessary phase. One set of arbitrary choices for the location of the zeros are -2.5 and -3.5. Of course, the zeros can also be located at the same location. There are unlimited number of choices. The proper choice for a particular design is determined by trial-and-error. That is, find the responses for a set of designs and select the one that gives the response closer to the desired one. The only suggestion is to place the compensator zero to the left of the desired poles, as the desired poles should be dominant. Once the location of a zero is fixed, the location of the corresponding pole can be fixed, as shown in earlier examples.

For this example, we get the transfer function of the lead part of the two-stage compensator as

$$K \frac{(s + 2.5)(s + 3.5)}{(s + 7.45)(s + 8.8627)}$$

With this compensator, the system response has to computed and verified that it satisfies the transient specifications. The gain constant K is determined as

$$|K| = \frac{|s_0 + 1| \times |s_0 + 2| \times |s_0 + 3| \times |s_0 + 7.45| \times |s_0 + 8.8627|}{|s_0 + 2.5| \times |s_0 + 3.5|}\Big|_{s_0 = -2 + j3.312}$$

$$= 158.1875$$

Next, the lag part of the compensator has to be determined. After some iterations, the transfer function of the lag part of the compensator is found to be

$$\frac{(s + 0.54)}{(s + 0.18)}$$

The gain and angle contribution of the lag compensator, for the present example, is 0.9578 and $-5.0007°$, respectively. Including that of the plant, we get the loop transfer function of the lead-lag compensated system as

$$158.1875 \frac{(s + 2.5)(s + 3.5)(s + 0.54)}{(s + 1)(s + 2)(s + 3)(s + 7.45)(s + 8.8627)(s + 0.18)}$$

$$= 158.1875 \frac{s^3 + 6.54s^2 + 11.99s + 4.7250}{s^6 + 22.4927s^5 + 178.9196s^4 + 613.0850s^3 + 928.8629s^2 + 544.5141s + 71.3093}$$

The root locus is shown in Fig. 7.24a. The desired poles have been changed slightly due to the phase of the lag compensator. The root locus segment between -3 and -2.5 and between 0.18 to 0.54 are not shown in the figure.

The corresponding closed-loop transfer function is

$$H(s) = \frac{158.2s^3 + 1035s^2 + 1897s + 747.4}{s^6 + 22.49s^5 + 178.9s^4 + 771.3s^3 + 1963s^2 + 2441s + 818.7}$$

The unit-step response of the compensated system is shown in Fig. 7.24b with performance figures.

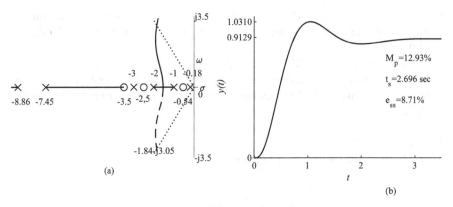

Fig. 7.24 (a) Root locus of $K\dfrac{s^3+6.54s^2+11.99s+4.7250}{s^6+22.4927s^5+178.9196s^4+613.0850s^3+928.8629s^2+544.5141s+71.3093}$; (b) unit-step response of $H(s)$

7.3 Summary

- There are three major tools used to easily analyze and design control systems, the root locus, Bode, and Nyquist plots. These tools bring out from the model some important characteristics of the system those can be used to analyze and design systems easily.
- The root locus of a system with characteristic equation $1 + G(s)F(s)$ are the loci of the of closed-loop poles as some parameter varies (typically, as the proportional gain varies from 0 to infinity).
- A plot of the points in the complex plane, in terms of real and imaginary parts of $s = \sigma + j\omega$, satisfying the angle condition alone is an alternative definition of the root locus.
- All the poles and zeros of a minimum-phase system lie in the left-half of the s-plane. Further, if both the system and its inverse are causal and stable, the system is of minimum-phase type.
- At least one pole or zero of a nonminimum-phase system lies in the right-half of the s-plane.
- A system associated with delay units is of nonminimum-phase type.
- Given $G(s)F(s)$, the design process of a control system has to ensure that the corresponding closed-loop system is stable with sufficient stability margin and has specified response.
- In a proportional compensator, the control signal at the output of the compensator is K_p times of its input.
- The control input, in a PI and lag compensator, has two components: (a) one is proportional to the error signal and (b) another is proportional to the integral of the error signal.

- The control input, in a PD or lead compensator, has two components: (a) one is proportional to the error signal and (b) another is proportional to the derivative of the error signal.
- The control input, in a PID or lag-lead compensator, has three components: (a) one is proportional to the error signal, (b) another is proportional to the integral of the error signal, and (c) yet another is proportional to the derivative of the error signal.

Exercises

* 7.1 Consider the loop transfer function

$$\frac{K}{(s+1)(s+3)}$$

of a unity feedback system. The design specification is that the overshoot M_p is to be limited to 3%. Use the proportional compensator to design the system.

7.2 Consider the loop transfer function

$$\frac{1}{(s+1)(s+3)}$$

of a unity feedback system. The design specification is that the overshoot of the unit-step response is to be limited to 8%. Further, the steady-state error is to be zero. Use the PI compensator to design the system.

* 7.3 Consider the loop transfer function

$$\frac{K}{(s+1)(s+3)}$$

of a unity feedback system. The design specification is that the overshoot is to be limited to 6%. Further, let the zero of the compensator z_c be 10 times that of the pole p_c. Use the lag compensator.

7.4 Consider the loop transfer function

$$\frac{K}{(s+1)(s+2)(s+3)}$$

of a unity feedback system. The design specification is that the overshoot is to be limited to 7%. Further, the settling time t_s is to be about 2.2 sec. Use the proportional-derivative compensator to design the system.

* 7.5 Consider the loop transfer function

$$\frac{K}{s(s+2)}$$

of a unity feedback system. The design specification is that the velocity error constant

$$K_v = \lim_{s \to 0} s G_c(s) L(s) \geq 20$$

The dimensionless damping ratio ζ is given as 0.5. Use the lead compensator to design the system.

7.6 Consider the loop transfer function

$$\frac{K}{(s+1)(s+2)(s+3)}$$

of a unity feedback system. The design specification is that the overshoot is to be limited to about 20% and settling time is to be about 2 s. Use the PID compensator to design the system.

* 7.7 Consider the loop transfer function

$$\frac{K}{(s+1)(s+2)(s+4)}$$

of a unity feedback system. The design specification is that the overshoot is to be limited to about 20%, the steady-state error is to be about 5% with respect to a constant reference and the settling time is to be about 2.5 s. Use the lag-lead compensator to design the system.

Chapter 8
Design of Control Systems in Frequency Domain: Bode Plot

Science and engineering students usually come across the frequency response for the first time in studying in such courses as signals and systems and audio systems. The concepts of frequency response, frequency range, and audible frequencies are critical for the understanding of audio systems. The frequency response is the system specification in several applications. The frequency-domain representation of signals and systems is, in every respect, just as complete and specific a representation as that in the time domain. The only difference is that the independent variable is frequency in the frequency domain in contrast to time in the time domain. Therefore, a control system can also be analyzed and designed in the frequency domain.

In solving simultaneous equations, if there are more variables than equations, we cannot find a unique solution, because there is not one. Similarly, in designing control systems, there are more parameters to fix than the available constraints. That is, we have to assume initial values for some parameters. Design the system and check the response. If the response is not satisfactory, redesign. This way, control system design is a trial-and-error procedure. Reducing the overshoot increases the rise time. Therefore, a compensator is required to have both smaller overshoot and faster rise time. The design of an appropriate compensator is the major task in control system design.

Some tasks are easier to carry out in the frequency domain and vice versa. For example, convolution operation in the time domain becomes much simpler multiplication in the frequency domain. Similarly, there are advantages and disadvantages of designing control systems in the two domains. The choice of the domain for the design has to be selected depending on the merits of the design for each specific case. Features of time-domain analysis and frequency-domain analysis are shown in Table 8.1.

The **frequency response** of a system is its response to sinusoids with infinite frequencies. Remember that the sinusoid is defined from $t = -\infty$ to $t = \infty$. Therefore, the sinusoidal response characterizes the steady-state behavior of the system. As the mathematically equivalent complex exponential representation of

© The Author(s), under exclusive license to Springer Nature Switzerland AG 2022
D. Sundararajan, *Control Systems*, https://doi.org/10.1007/978-3-030-98445-8_8

Table 8.1 Features of time-domain analysis and frequency-domain analysis

Time domain	Frequency domain
Analysis of higher-order systems difficult	Analysis of higher-order systems much easier
No unified method for design	Graphical method available for design
Mathematical model required	Readily measurable experimental data enough
Analysis of noise sensitivity difficult	Analysis of noise sensitivity easier

a real sinusoid is more convenient for analytical purposes, it is more often used in the analysis. The sinusoidal response $H(j\omega)$ is obtained by replacing the complex variable s by $j\omega$ in the Laplace-domain transfer function $G(s)$. While the response of practical systems is always real, due to the use of complex sinusoids, the response becomes complex-valued. The plot of the magnitude of the complex-valued frequency response versus frequency ω is called the **magnitude spectrum**. The plot of the phase angle of the complex-valued response versus frequency ω is called the **phase spectrum**. This form of the spectral plot is widely used and better understood, although the plots of real and imaginary parts versus the frequency ω could also be chosen. In the frequency response plots, the independent variable is, obviously, the frequency ω. Locus is a set of points representing a given condition. For example, the locus of points equidistant from a given point is a circle. A less familiar form is to plot the locus of all the points, as ω varies, representing real and imaginary components of the frequency response (or the magnitude and phase components). It is a rectangular plot with coordinates real and imaginary (or magnitude and phase) components with frequency ω as a parameter. We get two interpretations with a single plot. As the response of practical systems is real-valued and each real value requires the values of two complex-conjugate values, the complex response is redundant by a factor of 2. Therefore, the plot over the frequency range $0 \le \omega \le \infty$ is adequate most of the time.

The frequency response of a linear time-invariant system is unique. But, in the Bode and Nyquist plots, the response is plotted in different formats. This difference in formats makes significant differences. A sufficiently accurate Bode plot can be drawn easily from the asymptotic response of poles and zeros by hand. Gain and phase margins can be more easily determined. The changes in the response in adding controllers or compensators and varying their parameters can be easily visualized. The limitation of the Bode plot is that the stability of minimum-phase systems only can be determined. Expect at the origin, a minimum-phase system has poles and zeros located only in the left-half of the s-plane. The magnitude and phase angle characteristics of a minimum-phase system are directly related. From the magnitude plot over the entire range of frequency, the phase plot can be uniquely determined and vice versa. As frequency approaches ∞, the slope of the magnitude plot of minimum-phase systems is ± 20 dB/decade for simple zeros and poles, respectively. The phase angle at $\omega = \infty$ is $\pm 90°$ for simple zeros and poles, respectively.

The frequency response characterizes a system in the frequency domain, which is the response to sinusoids. The frequency response of a system can be easily determined in the laboratory by readily available equipments. The essential resources used in the analysis and design are graphical, the Bode and Nyquist plots. From these plots, the absolute and relative stabilities of linear closed-loop systems are determined from the knowledge of their open-loop frequency response. The basic objective is to ensure the stability of the system with adequate performance.

In closed-loop control systems, part of the output is fed back to reduce the system error. If the feedback is negative, it helps to reduce the error. But, positive feedback increases the error. The transient response terms must be decaying exponentials. That is, all the roots of the characteristic equation must be negative. The response in the frequency domain has magnitude and phase components. The negative phase due to the poles increases with frequency. At some frequency ω, the phase becomes $-180°$. The negative feedback becomes positive feedback, which has to be avoided. Therefore, it has to be ensured that the phase response is less than $-180°$ by proper design with sufficient gain. This requires compensating networks to correct the phase response suitably.

8.1 Bode Plot

Bode plot is a frequency response plot, which is often used to study the characteristics of the loop transfer function of a feedback system. The loop transfer function of the system is $G(s)F(s)$. The frequency response is obtained as $G(j\omega)F(j\omega)$ by replacing s by $j\omega$. The frequency response can be easily obtained using an oscillator to provide the sinusoidal input of frequencies of interest, applying the sinusoids to the system and measuring the change in the magnitude and phase of the input sinusoids at the output. A large loop gain improves the performance of a system. However, it could also lead to instability of the system. Bode plot is one of the methods to test the relative stability of a system.

The two plots of the logarithmic magnitude in decibels and phase responses of a system, with logarithmic frequency scales, to sinusoidal inputs are known as **Bode** plots. These plots can be constructed easily using the asymptotic behavior of the responses and are used widely in practical feedback system analysis and design. Consider the loop transfer function of unity feedback system with its numerator and denominator polynomials in factored form

$$G(s) = \frac{K(s + z_1)(s + z_2)}{s(s + p_1)(s^2 + p_{21}s + p_{22})}$$

where K is a positive constant. The second-order term in the denominator is assumed to have complex-conjugate roots. Now, we replace s by $j\omega$. Furthermore, the transfer function is rearranged in a normalized form to facilitate plotting the Bode plot.

$$G(j\omega) = \frac{\left(\frac{Kz_1z_1}{p_1p_{22}}\right)\left(\frac{j\omega}{z_1} + 1\right)\left(\frac{j\omega}{z_2} + 1\right)}{j\omega\left(\frac{j\omega}{p_1} + 1\right)\left(\frac{(j\omega)^2}{p_{22}} + \frac{(j\omega)p_{21}}{p_{22}} + 1\right)}$$

The zero-order term of the factors has been reduced to unity if it exists. Multiplication and division operations are required to find the magnitude of the total response. In order to use addition and subtraction operations to compute both the total magnitude and phase, the magnitude response in dB and the phase response are, respectively, expressed as

$$20 \log_{10} |G(j\omega)|$$

$$= 20 \log_{10} \left|\left(\frac{Kz_1z_1}{p_1p_{22}}\right)\right| + 20 \log_{10} \left|\left(\frac{j\omega}{z_1} + 1\right)\right| + 20 \log_{10} \left|\left(\frac{j\omega}{z_2} + 1\right)\right|$$

$$-20 \log_{10} |j\omega| - 20 \log_{10} \left|\left(\frac{j\omega}{p_1} + 1\right)\right|$$

$$-20 \log_{10} \left|\left(\frac{(j\omega)^2}{p_{22}} + \frac{(j\omega)p_{21}}{p_{22}} + 1\right)\right|$$

$$\angle G(j\omega) = \angle\left(\frac{j\omega}{z_1} + 1\right) + \angle\left(\frac{j\omega}{z_2} + 1\right) - \angle j\omega - \angle\left(\frac{j\omega}{p_1} + 1\right)$$

$$-\angle\left(\frac{(j\omega)^2}{p_{22}} + \frac{(j\omega)p_{21}}{p_{22}} + 1\right)$$

The corner frequencies of the two zeros are z_1 and z_2. At a **corner frequency**, the slope of the asymptotic approximation of the frequency response changes or the frequency at which the two magnitude asymptotes intersect. The corner frequencies of the first- and the second-order poles are p_1 and $\sqrt{p_{22}}$. The basic factors of the transfer function are:

1. A constant term
2. A pole or zero at the origin
3. A first-order pole or zero
4. A second-order pole or zero

The Constant
The magnitude of a constant K in decibels is $20 \log_{10} |K|$. If the constant is positive, its phase is zero degrees and it is 180° for a negative constant. The value is positive for a K greater than unity and it is negative for a K less than unity. The Bode plot for a constant is a straight line. Figure 8.1a and b show the magnitude and phase plots, respectively, for $K = 10, 0.1, -\sqrt{2}$.

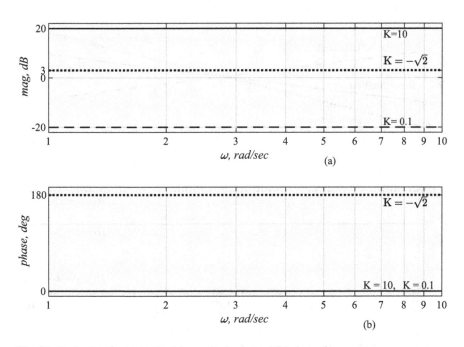

Fig. 8.1 Bode plots for constants: (**a**) magnitude plot and (**b**) phase plot

$$20\log_{10}(10) = 20 \text{ dB}, \ 20\log_{10}(0.1) = -20\log_{10}(10) =$$
$$- 20 \text{ dB}, \ 20\log_{10}(|-\sqrt{2}|) = 3.0103 \text{ dB}$$

Pole or Zero at the Origin

The magnitude of a zero $j\omega$ is ω and its phase is a constant $90°$. The function ω is a linear ramp and it is 10 at $\omega = 10$ and it is 100 at $\omega = 100$. The corresponding values in decibels are $20\log_{10}(10) = 20$ dB and 40 dB. Therefore, the plot of a zero has a slope of 20 dB per decade (as the frequency changes 10 times). The slope is $20\log_{10}(2) = 6.0206 \approx 6$ dB per octave (as the frequency doubles). The slope of a second-order zero $(j\omega)^2$ is $2 \times 20 = 40$ dB per decade or 12 dB per octave. The phase is a constant $180°$. The plot is a linear ramp with appropriate slope with value 0 dB at $\omega = 1$.

For a pole, the slope of the magnitude and the phase have the same magnitude of a zero but are negative. The plots are mirror images of each other about the ω axis. In general, for $(j\omega)^{\pm n}$, the magnitude is $\pm n(20\log_{10}(\omega))$ and the phase is $\pm n(90°)$. Figure 8.2a and b show the magnitude and phase plots, respectively, for

$$H(j\omega) = j\omega, \quad H(j\omega) = (j\omega)^2, \quad H(j\omega) = \frac{1}{j\omega}$$

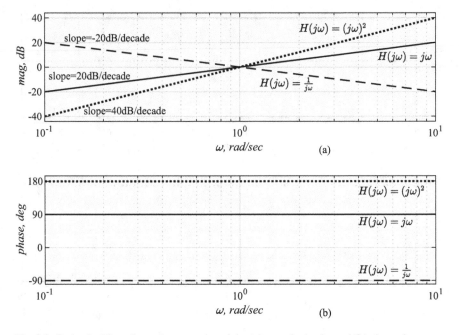

Fig. 8.2 Bode plots for poles and zeros at the origin: (**a**) magnitude plot and (**b**) phase plot

First-Order Pole or Zero

The magnitude in decibels of a first-order zero at $-z_1$ is $20 \log_{10} |(\frac{j\omega}{z_1} + 1)|$. For $\omega \ll z_1$,

$$20 \log_{10} |\left(\frac{j\omega}{z_1} + 1\right)| \approx 20 \log_{10}(1) = 0$$

Therefore, the log magnitude tends to zero asymptotically for $\omega \ll z_1$. For $\omega \gg z_1$,

$$20 \log_{10} \left|\left(\frac{j\omega}{z_1} + 1\right)\right| = 20 \log_{10} \left(\frac{\omega}{z_1}\right) = -20 \log_{10}(z_1) + 20 \log_{10}(\omega)$$

Therefore, the log magnitude plot is a ramp with slope 20 dB/decade or 6 dB/octave. The two asymptotes meet at $\omega = z_1$, called the corner frequency. When $\omega = z_1$, the magnitude is zero.

The plot for a pole is the mirror image, about the frequency axis, of that of a zero at the same corner frequency. The phase of a zero is 0 for $\omega \ll z_1$ and it is 90° for $\omega \gg z_1$. Figure 8.3a and b show the magnitude and phase Bode plots, respectively, for

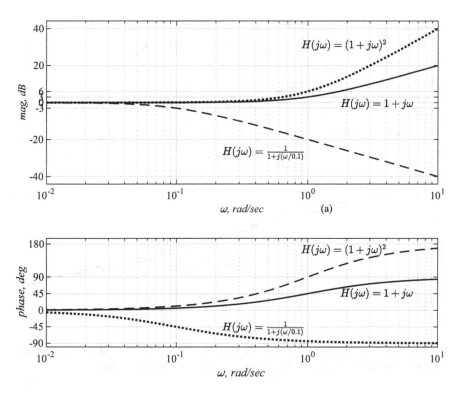

Fig. 8.3 Bode plots for first-order poles and zeros: (**a**) magnitude plot and (**b**) phase plot

$$H(j\omega) = 1 + j\omega, \quad H(j\omega) = (1 + j\omega)^2, \quad H(j\omega) = \frac{1}{1 + j(\omega/0.1)}$$

The exact magnitude of the response for a zero is

$$20 \log_{10} \left| \left(\frac{j\omega}{z_1} + 1 \right) \right| = 20 \log_{10} \sqrt{\left(\frac{\omega^2}{z_1^2} + 1 \right)} = 10 \log_{10} \left(\frac{\omega^2}{z_1^2} + 1 \right)$$

For a pole, it is the mirror image, about the frequency axis, of that of a zero at the same corner frequency. As shown in Fig. 8.4a, the maximum difference between exact and asymptotic magnitude responses is 3 dB at $\omega = z_1$, the corner frequency. The error is 1 dB at one octave above and below the corner frequency. The exact plot can be approximated closely by adding these differences to the asymptotic (a straight line that is a limiting value of a curve can be considered as tangent at infinity) plots. The corresponding exact and asymptotic phase plots are shown in Fig. 8.4b.

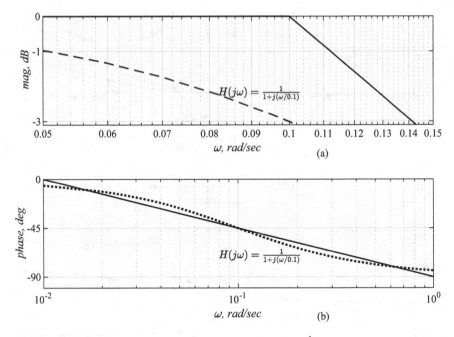

Fig. 8.4 Bode plots, asymptotic and exact, for first-order pole $\frac{1}{1+j(\omega/0.1)}$ near the corner point: (**a**) magnitude plot and (**b**) phase plot. The asymptotic plots are shown by solid lines

Quadratic Factors

In control system applications, a second-order transfer function is written in the form

$$G(j\omega) = \frac{1}{1 + 2\zeta(j\frac{\omega}{\omega_n}) + (j\frac{\omega}{\omega_n})^2}$$

where ζ is the damping ratio and ω_n is the undamped natural frequency. If $\zeta > 1$, the pole can be decomposed into a product of two first-order poles located on the real axis of the complex plane. If $0 < \zeta < 1$, then the two first-order poles form a complex-conjugate pair. For low values of ζ, the approximation of the frequency response by asymptotes becomes less accurate, as the response is a function of both the corner frequency and ζ.

The corresponding log magnitude and phase expressions are

$$20\log_{10}|G(j\omega)| = -20\log_{10}\left|1 + 2j\zeta\frac{\omega}{\omega_n} + \left(j\frac{\omega}{\omega_n}\right)^2\right|$$

and

$$\angle(G(j\omega)) = -\angle\left(1 + 2j\zeta\frac{\omega}{\omega_n} + \left(j\frac{\omega}{\omega_n}\right)^2\right) = -\tan^{-1}\left(\frac{2\zeta\left(\frac{\omega}{\omega_n}\right)}{1 - \left(\frac{\omega}{\omega_n}\right)^2}\right)$$

For $\omega \ll \omega_n$,

$$20\log_{10}|H(j\omega)| \approx -20\log_{10}(1) = 0$$

For $\omega \gg \omega_n$,

$$20\log_{10}|H(j\omega)| \approx -20\log_{10}\left|\left(-\frac{\omega}{\omega_n}\right)^2\right| = -40\log_{10}\left(\frac{\omega}{\omega_n}\right)$$

$$= -40\log_{10}\omega + 40\log_{10}\omega_n$$

At $\omega = \omega_n$, the magnitude is approximated to zero and the corner frequency is ω_n. High frequency asymptote has a slope of -40 dB/decade. The exact magnitude is given by

$$20\log_{10}|H(j\omega)| = -20\log_{10}\sqrt{\left(1 - \left(\frac{\omega}{\omega_n}\right)^2\right)^2 + \left(2\zeta\frac{\omega}{\omega_n}\right)^2}$$

The exact phase response is given by

$$\angle(H(j\omega)) = -\tan^{-1}\left(\frac{2\zeta(\frac{\omega}{\omega_n})}{1 - (\frac{\omega}{\omega_n})^2}\right)$$

At $\omega = 0$, the phase angle is 0. At $\omega = \infty$, the phase angle is $-180°$. At $\omega = \omega_n$, the phase angle is $-90°$. For complex-conjugate zeros, the plots are the mirror images of those for complex-conjugate poles. Figure 8.5a and b show the magnitude and phase Bode plots for

$$G(j\omega) = \frac{1}{1 + j\omega + (j\omega)^2} \quad \text{and} \quad G(j\omega) = \frac{1}{1 + j2\omega + (j\omega)^2}$$

The resonant frequency ω_r is

$$\omega_r = \omega_n\sqrt{1 - 2\zeta^2}, \quad 0 \leq \zeta \leq 0.707$$

It is evident that $\omega_r \to \omega_n$ as $\zeta \to 0$. The magnitude at ω_r is

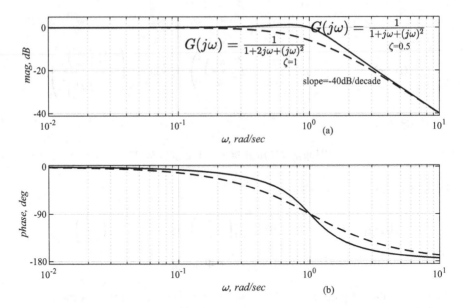

Fig. 8.5 Bode plot for quadratic transfer functions with damping ratio 0.5 and 1: (**a**) magnitude plot and (**b**) phase plot

$$M_r = \frac{1}{2\zeta\sqrt{1-\zeta^2}}$$

and for $\zeta > 0.707$, $M_r = 1$. For $\zeta = 0.5$, $M_r = 1.1547$.

As always, although software packages are used for plotting Bode plot, it is essential to do some plots manually for good understanding of the theory and for verification. First, decompose the transfer function as a product of basic factors. Then, compute the magnitude and phase response for all the basic factors individually. Then, sum all of them to get the complete plot. The easiness of plotting and modification of Bode plots, as required, is one of the major reasons for its widespread use in practice.

Example 8.1 Plot the Bode plot of a system with the transfer function $G(s)$.

$$G(s) = \frac{10s}{(s+1)(s+3)(s^2+s+1)}$$

Determine the gain and margins of the system.

Solution Writing the transfer function in normalized form, we get

$$G(s) = \frac{\frac{10}{3}s}{(\frac{s}{1}+1)(\frac{s}{3}+1)(s^2+s+1)}$$

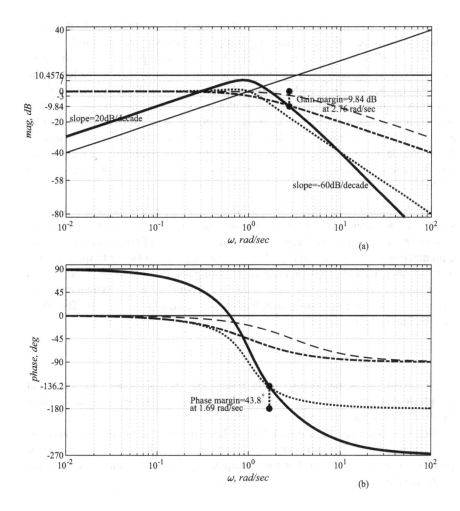

Fig. 8.6 Bode plot for $G(s)$ and its factors: (**a**) magnitude plot and (**b**) phase plot

Figure 8.6a and b show the magnitude and phase Bode plots, respectively, for $G(j\omega)$ and its factors. The magnitude and phase angle plots of the constant 10/3 are, respectively, horizontal lines with a value of 10.4576 dB and 0°, shown by solid lines. The magnitude and phase angle plots of the zero at the origin are, respectively, a straight line with a slope of 20 dB/decade passing through the point $\omega = 1$ and a horizontal straight line with a value of 90°, shown by a solid line. The magnitude plot of the pole $(s + 1)$ is, respectively, a horizontal line with a value of 0 dB to the left of the corner frequency at $\omega = 1$ and a ramp with a slope of -20 dB/decade to the right, shown by a dash–dot line. The phase angle plot has values of 0, -45, and $-90°$ at $\omega = 0$, $\omega = 1$, and $\omega = \infty$, respectively. The magnitude plot of the pole $(\frac{s}{3}s + 1)$ is, respectively, a horizontal line with a value of 0 dB to the left of the corner frequency at $\omega = 3$ and a ramp with a slope of -20 dB/decade to the right, shown by a dashed line. The phase angle plot has values of 0, -45, and $-90°$ at $\omega = 0$, $\omega = 3$, and $\omega = \infty$, respectively. The magnitude plot of the double pole $(s^2 + s + 1)$ is, respectively, a horizontal line with a value of 0 dB to the left of the corner frequency at $\omega = 1$ and a ramp with a slope of -40 dB/decade to the right, shown by a dotted line. The phase angle plot has values of 0, -90, and $-180°$ at $\omega = 0$, $\omega = 1$, and $\omega = \infty$, respectively. The complete plot for $G(s)$ is shown by a thick line. It is the sum of the plots of all its component factors. It has a slope of 20 dB/decade to the left of $\omega = 1$ and a slope of -60 dB/decade to the right. It has a peak value of 7 dB at $\omega = 1$. The phase starts with a value of $90°$ and ends at $-270°$.

The closeness of the $G(j\omega)$ locus to the $(-1, j0)$ point is an indicator of the stability of a system. This requires two measures in the Bode plots. The **phase crossover frequency** is the frequency at which the phase of $G(j\omega)$ is $-180°$. The **gain margin** is defined as $-20 \log 10 |G(j\omega)|$ at the phase crossover frequency. That is, it indicates the additional gain required to bring a stable minimum-phase system to the verge of instability. The **gain crossover frequency** is the frequency ω_g, at which the gain of the open-loop transfer function is equal to 1 or 0 dB. The phase margin, Pm, is defined as

$$\text{Pm} = \angle(G(j\omega_g)) - 180°$$

where $\angle(G(j\omega_g))$ is the positive angle in degrees. An equivalent definition is that the **phase margin** is $180° + \angle(G(j\omega_g))$ at the gain crossover frequency, where $\angle(G(j\omega_g))$ is negative.

That is, it indicates the additional phase lag required to bring the system to the verge of instability. For minimum-phase systems, both the measures have to be positive for a stable system. For the example, Fig. 8.6a and b show the gain and phase margins.

8.1.1   Bode Plot of a Lag Compensator

The lag compensator is of the form

$$G_c(s) = \frac{s + z}{s + p}, \qquad |p| < |z|$$

The low frequency gain is

$$\frac{z}{p}$$

and 0 dB at high frequencies. The lag compensator is essentially a lowpass filter, suppressing high frequency components. It improves the steady-state error. However, the phase margin is reduced due to the negative phase. This problem is reduced by the zero component of the compensator placed appropriately. That is, lag compensator adds gain at low frequencies without affecting phase margin. The phase is

$$\angle(G_c(s)) = \angle(s+z) - \angle(s+p)$$

Letting $s = j\omega$, we get

$$\angle(G_c(j\omega)) = \angle(j\omega + z) - \angle(j\omega + p) = \tan^{-1}\left(\frac{\omega}{z}\right) - \tan^{-1}\left(\frac{\omega}{p}\right)$$

The derivative of $\tan^{-1}(ax)$ with respect to x is $\frac{a}{1+(ax)^2}$. Let us find the frequency ω_{max} where the maximum phase lag occurs for the case $p = 0.1$ and $z = 1$. Differentiating the expression for phase with respect to ω and equating it to zero, we get

$$\frac{1}{1+\omega^2} - \frac{10}{1+100\omega^2} = 0$$

Solving for ω, we get $\omega = 0.3162$ rad/sec. Looking at the figure, the maximum phase lag occurs at the center of the corner frequencies of the pole and zero, which is the geometric mean of p and z. That is, $\omega_{max} = \sqrt{zp}$. For this case, $\omega_{max} = \sqrt{1 \times 0.1} = 0.3162$, as found earlier. The maximum phase lag is

$$\tan^{-1}\left(\frac{0.3162}{1}\right) - \tan^{-1}\left(\frac{0.3162}{0.1}\right) = -54.9032°$$

Similarly, for the case $p = 0.2$ and $z = 1$, $\omega_{max} = \sqrt{1 \times 0.2} = 0.4472$ rad/sec and maximum phase lag is $-41.8103°$. The Bode plot of the lag compensator with the transfer functions

$$G_c(s) = \frac{s+1}{s+0.1} \quad \text{and} \quad G_c(s) = \frac{s+1}{s+0.2}$$

is shown in Fig. 8.7, respectively, by solid and dashed lines. The low frequency gain with pole at $p = 0.1$ is

$$20\log 10\left(\frac{z}{p}\right) = 20\log 10(10) = 20\,\text{dB}$$

and 0 dB at high frequencies. The low frequency gain with pole at $p = 0.2$ is

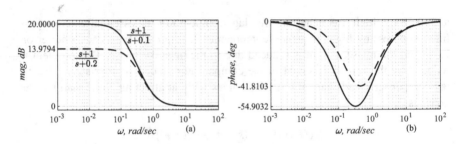

Fig. 8.7 Bode plot of lag compensators: (**a**) magnitude and (**b**) phase

$$20 \log 10 \left(\frac{z}{p} \right) = 20 \log 10(5) = 13.9794 \, \text{dB}$$

and 0 dB at high frequencies.

8.1.2   Bode Plot of a Lead Compensator

Lead compensator adds gain at higher frequencies. It is basically a highpass filter. The lead compensator is of the form

$$G_c(s) = \left(\frac{p_{lead}}{z_{lead}} \right) \left(\frac{s + z_{lead}}{s + p_{lead}} \right), \qquad p_{lead} > z_{lead}$$

At very low frequencies, both the gain and phase tend to zero. At very high frequencies, the phase tends to zero and the gain tends to p_{lead}/z_{lead}. The phase is

$$\angle(G_c(s)) = \angle(s + z) - \angle(s + p)$$

Letting $s = j\omega$, we get

$$\angle(G_c(j\omega)) = \angle(j\omega + z) - \angle(j\omega + p) = \tan^{-1} \left(\frac{\omega}{z} \right) - \tan^{-1} \left(\frac{\omega}{p} \right)$$

The maximum phase lead of the compensator is θ_{max} and it occurs at ω_{max}. Looking at the figure, the maximum phase lead occurs at the center of the corner frequencies of the pole and zero, which is the geometric mean of p and z; that is, $\omega_{max} = \sqrt{zp}$. The Bode plot of the lead compensator with the transfer functions

$$G_c(s) = 10 \frac{s + 0.1}{s + 1} \quad \text{and} \quad G_c(s) = 5 \frac{s + 0.2}{s + 1}$$

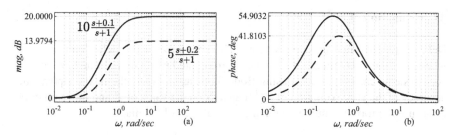

Fig. 8.8 Bode plot of lead compensators: (**a**) magnitude and (**b**) phase

is shown in Fig. 8.8, respectively, by solid and dashed lines. The low frequency gain with pole at $p = 0.1$ is

$$20 \log 10 \left(\frac{p}{z} \frac{z}{p} \right) = 20 \log 10(1) = 0 \, \text{dB}$$

and the gain is 20 dB at high frequencies. The low frequency gain with pole at $p = 0.2$ is

$$20 \log 10 \left(\frac{p}{z} \frac{z}{p} \right) = 20 \log 10(1) = 0 \, \text{dB}$$

and the gain is 13.9794 dB at high frequencies. The maximum phase lead provided by a single stage compensator is 65° in practice. For more phase lead, compensator units can be cascaded.

Let $\beta = z_{lead}/p_{lead}$, $\beta < 1$.

$$\theta_{max} = \sin^{-1} \left(\frac{1 - \beta}{1 + \beta} \right)$$

The parameter β can be determined for a desired θ_{max}; that is,

$$\beta = \frac{(1 - \sin(\theta_{max}))}{(1 + \sin(\theta_{max}))}$$

The phase at ω_{max} is θ_{max}. Then,

$$z\_lead = \sqrt{\beta}\omega_{max}, \qquad p\_lead = z\_lead/\beta$$

In the design of systems with lead compensators, these formulas are used.

Approximation of e^{-Ts}

Delay units are often used in systems. The transform of the delay unit is e^{-Ts}, where T is the delay time. This can be approximated by a rational function, so that such systems can be analyzed using Bode plots, as shown in the next chapter. The accuracy of the approximation of e^{-Ts} depends on the value of T and the number of terms used to form the numerator and denominator polynomials.

8.2 Design of Control Systems

The basic steps in the design of control systems are:

1. Get the specification
2. Determine the type of controller or compensator structure and the way it is to be connected to the controlled process
3. Determine the parameters of the controller so that the given specifications are met with sufficient stability margin

The specifications include transient and frequency responses, relative stability, and sensitivity to parameter variations. A common type of controller, called PID controller, uses some combination of the proportional, integral, and derivative of the error signal. Another type, called lead, lag, and lead–lag controllers, is also often used, which provides various types of compensation.

In the last chapter, we used the root locus of the loop transfer function as a tool to design various types of control systems. In this chapter, we use the frequency response of the loop transfer function as a tool to design various types of control systems. Both methods should be tried and take the design that yields the better response. The types of systems and the design specifications essentially remain the same.

In the frequency-domain design, we control the damping ratio ζ using the fact that the phase margin is a function of ζ. Steady-state error is controlled by the gain of the loop frequency response as $s \rightarrow 0$. Settling time decreases with increasing bandwidth. In turn, the bandwidth is determined by the compensator. It is expected since higher frequency components of the frequency response control the speed of response. Furthermore, the key parameters in the frequency-domain are the gain margin, the phase margin, resonant frequency and magnitude, gain and phase crossover frequencies. Iterations are unavoidable in the design. The first step is to make an initial design using the given specifications. Then, depending on the shortcomings of the response, we have to make a redesign by changing the poles and zeros of the compensators and gain until the response of the compensated system is satisfactory with good stability margin. With sufficient practice, we can get used to the design procedure.

8.2.1 Relation Between Time-Domain and Frequency-Domain Specifications

A second-order closed-loop transfer function is written in the standard form as

$$H(s) = \frac{\omega_n^2}{(s^2 + 2\zeta\omega_n s + \omega_n^2)}$$

where ζ is the damping ratio and ω_n is the undamped natural frequency. At sinusoidal steady-state, with $s = j\omega$, the transfer function can be equivalently written as

$$H(j\omega) = \frac{1}{1 + j2\zeta\left(\frac{\omega}{\omega_n}\right) - \left(\frac{\omega}{\omega_n}\right)^2}$$

Let $v = (\omega/\omega_n)$. Then,

$$H(jv) = \frac{1}{1 + j2\zeta v - v^2}$$

The bandwidth, BW, is the frequency at which the magnitude of the frequency response drops to 0.707 (or 3 dB) of its value at $\omega = 0$. Then, the magnitude is

$$|H(jv)| = \frac{1}{\sqrt{(1 - v^2)^2 + (2\zeta v)^2}} = \frac{1}{\sqrt{2}} \quad \text{or} \quad (1 - v^2)^2 + (2\zeta v)^2 = 2$$

Solving for v^2, we get

$$v^2 = (1 - 2\zeta^2) + \sqrt{4\zeta^4 - 4\zeta^2 + 2},$$

since v must be a positive real quantity for any ζ. The **bandwidth** of $H(jv)$ is given by

$$\text{BW} = \omega_n\sqrt{(1 - 2\zeta^2) + \sqrt{4\zeta^4 - 4\zeta^2 + 2}}$$

A higher bandwidth results in a faster response. Increasing ω_n increases bandwidth and increasing ζ decreases bandwidth.

Differentiating $|H(jv)|$ with respect to v and equating it to zero, we get the resonant frequency ω_r as

$$v_r = \sqrt{1 - 2\zeta^2} \quad \text{or} \quad \omega_r = \omega_n\sqrt{1 - 2\zeta^2}$$

Substituting for ω_r in the expression for $|H(jv)|$, we get

$$M_r = \frac{1}{2\zeta\sqrt{1-\zeta^2}}, \qquad \zeta \le \frac{1}{\sqrt{2}}$$

For $\zeta > 0.707$, $\omega_r = 0$ and $M_r = 1$.

Relation Between Phase Margin and the Damping Ratio ζ

A second-order loop transfer function with unity feedback is written in the standard form as

$$G(s) = \frac{\omega_n^2}{s(s+2\zeta\omega_n)}$$

where ζ is the damping ratio and ω_n is the undamped natural frequency. Equating the magnitude of $G(s)$ to unity and solving for the frequency of its occurrence, we get the loop gain crossover frequency as

$$\omega_{gc} = \omega_n\sqrt{\sqrt{(1+4\zeta^4)}-2\zeta^2}$$

$$\text{Phase margin} = 180 + \angle(Gj\omega_{gc}) = \tan^{-1}\frac{2\zeta}{\sqrt{\sqrt{(1+4\zeta^4)}-2\zeta^2}}$$

Figure 8.9 shows the relationship between the phase margin and the damping ratio ζ by a solid line. They are directly related. The relationship is approximately linear in the range $0 \le \zeta \le 0.6$, as shown by the dashed line; that is,

$$\text{phase margin} \approx 100\zeta$$

This relationship is often used in the frequency-domain design of control systems. In general, adding a zero to $G(s)$ increases the bandwidth of the closed-loop system, while adding a pole to $G(s)$ decreases the bandwidth (making the corresponding closed-loop system less stable).

Let us recollect the definitions of the position, velocity and acceleration error constants, and the steady-state errors.

$$K_p = \lim_{s\to 0} G(s), \qquad K_v = \lim_{s\to 0} sG(s), \qquad K_a = \lim_{s\to 0} s^2G(s)$$

For an input with magnitude R, the errors are

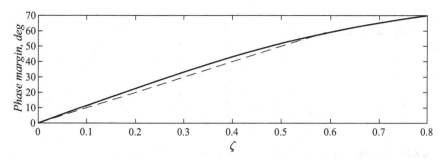

Fig. 8.9 Phase margin versus the damping ratio ζ

$$e_{ss} = \frac{R}{1 + K_p}, \qquad e_{ss} = \frac{R}{K_v}, \qquad e_{ss} = \frac{R}{K_a}$$

8.2.2 Lag Compensator

Example 8.2 Design a lag compensator for the system with the loop transfer function

$$G(s) = \frac{1}{(s + 1)(s + 2)}$$

of a unity feedback system. The specifications are (i) steady-state error $e_{ss} < 3\%$ for step input and (b) the overshoot is to be limited to 15%.

Solution With maximum overshoot $os = 0.15$,

$$\zeta = -\frac{\log_e(os)}{\sqrt{\pi^2 + (\log_e(os))^2}} = 0.5169$$

The required phase margin to satisfy the overshoot specification is

$$Pm \approx 100\zeta = 51.69°$$

Adding 10° for compensator phase, the required phase margin at the gain crossover frequency ω_{Pm} is 61.6931°.

Step 1 The gain of the uncompensated system is to be adjusted to give the required phase margin 61.6931°. The Bode plot of the uncompensated system is shown in Fig. 8.10 by a dashed line. Locate the frequency at which the phase is $-180 + 61.6931° = -118.3069°$. The frequency is $\omega = 2.437$ and the gain is -18.3875 dB, as shown in the figure. To make 61.6931° as the phase margin, the corresponding gain must be 1 or zero dB. Therefore, the gain has to be increased by 18.3875 dB.

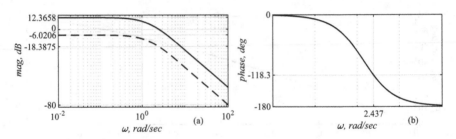

Fig. 8.10 Bode plot of $G(s) = \frac{1}{(s+1)(s+2)}$: (**a**) magnitude and (**b**) phase

The magnitude of the Bode plot of the gain compensated system, with the transfer function

$$KG(s) = \frac{8.3046}{(s+1)(s+2)},$$

is shown in Fig. 8.10a by a solid line.

Step 2 Determine the static error constant K_p from Fig. 8.10. This constant is just the DC gain of 12.3658 dB, which corresponds to $K_p = 10^{12.3658/20} = 4.1523$. Now, from the specified position error, the required error constant is

$$K_{pr} = \frac{1}{e_{ss}} - 1 = \frac{1}{0.03} - 1 = 32.3333$$

The error constant has to be improved by a factor $\alpha = 32.3333/4.1523 = 7.7868$.

Step 3 Design the compensator. Locate the corner frequency of the zero of the compensator one decade below the gain crossover frequency, $2.437/10 = 0.2437$ rad/sec. The zero of the compensator is $s + 0.2437$. The zero reduces the phase at gain crossover frequency. The pole of the compensator is $s + 0.2437/\alpha = s + 0.0313$. The transfer function of the compensator is

$$G_c(s) = \frac{(s + 0.2437)}{(s + 0.0313)}$$

The lag compensator adds gain in the low frequency part of the Bode plot only and reduces the steady-state error. The transfer function of the lag-compensated system is

$$KG_c(s)G(s) = \frac{(8.305s + 2.024)}{(s^3 + 3.031s^2 + 2.094s + 0.06259)}$$

The Bode plot of the transfer function is shown in Fig. 8.11, along with its components.

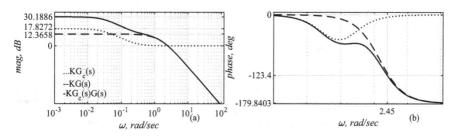

Fig. 8.11 Bode plot of $KG_c(s)G(s) = \frac{(8.305s+2.024)}{(s^3+3.031s^2+2.094s+0.06259)}$ and its components: (**a**) magnitude and (**b**) phase

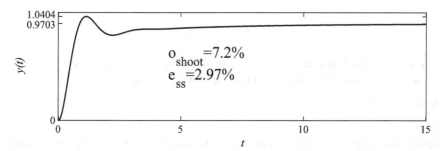

Fig. 8.12 Unit-step response of $H(s) = \frac{(8.305s+2.024)}{(s^3+3.031s^2+10.4s+2.086)}$

The corresponding closed-loop transfer function is

$$H(s) = \frac{(8.305s + 2.024)}{(s^3 + 3.031s^2 + 10.4s + 2.086)}$$

The unit-step response is shown in Fig. 8.12. The response meets the specifications. However, the settling time is very long.

8.2.3 Lead Compensator

There are three key parameters in the lead compensator design. The gain crossover frequency ω_{Pm} determines the closed-loop bandwidth ω_{BW}, the rise time t_r, the peak time t_p, and the settling time t_s. The phase margin Pm determines the damping ratio ζ and overshoot. The low frequency gain determines the steady-state performance.

Example 8.3 Design a lead compensator for the system with the open-loop transfer function

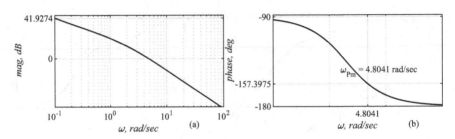

Fig. 8.13 Bode plot of $G(s) = \frac{25}{s(s+2)}$: (a) magnitude and (b) phase

$$G(s) = \frac{1}{s(s+2)}$$

of a unity feedback system. The specifications are (i) steady-state error $e_{ss} < 10\%$ for ramp input and (b) the overshoot is to be limited to 15%.

Solution

Step 1
Set the loop-gain K of $G(s)$ to provide the required static error constant and find the corresponding phase margin with this K.

For the given system, steady-state error will be 0.1, if the system gain is increased to 20; that is,

$$K_v = \lim_{s \to 0} \frac{20s}{s(s+2)} = 10$$

For a conservative design, let us make the gain 25 and the transfer function becomes

$$G(s) = \frac{25}{s(s+2)}$$

The Bode plot of $G(s)$ is shown in Fig. 8.13. The phase margin is 22.6025°.

Step 2
Compute ζ from overshoot specifications and determine the required phase margin. Find the required phase lead to be provided by the compensator.

Although the specified overshoot is 15%, let us design for a conservative 13%. Then, we get

$$\zeta = -\frac{\log_e(os)}{\sqrt{\pi^2 + (\log_e(os))^2}} = 0.5446$$

Setting the phase margin Pm as

$$Pm = 100\zeta + 10° = 64.4648°,$$

where the additional $10°$ is to offset the shifting of the gain crossover frequency to the right by the addition of the compensator. The required phase lead is

$$\theta_{max} = 64.4648° - 22.6025° = 41.8623°$$

Step 3
Compute β, the gain, the zero z_lead, and the pole p_lead of the compensator.
Using θ_{max}, determine β as

$$\beta = \frac{(1 - \sin(\theta_{max}))}{(1 + \sin(\theta_{max}))} = 0.1995$$

Setting $\omega_{max} = \omega_{Pm} = 4.8041$,

$$z\_lead = \omega_{max}\sqrt{\beta} = 2.1458, \qquad p\_lead = z\_lead/\beta = 10.7554$$

The gain K_c of the compensator is

$$K_c = \frac{1}{\beta} = 5.0122$$

Step 4
Verify that the response of the lead-compensated system meets the specifications, otherwise iterate.
The transfer function of the lead compensator is

$$G_c(s) = 5.0122\frac{s + 2.1458}{s + 10.7554}$$

The transfer function of the lead-compensated system is

$$G_c(s)G(s) = 5.0122\left(\frac{s + 2.1458}{s + 10.7554}\right)\left(\frac{25}{s(s + 2)}\right)$$

$$= 125.305\frac{s + 2.1458}{s^3 + 12.7554s^2 + 21.5108s}$$

While the design satisfies the steady-state error specification, the overshoot is more than desired. Then, the first choice is to try to solve the problem by gain adjustment. By trial-and-error, the overall gain of 100.2 has been found to give the desired response; that is, the transfer function of the lead-compensated system is

$$G_c(s)G(s) = 100.2\frac{s + 2.1458}{s^3 + 12.7554s^2 + 21.5108s}$$

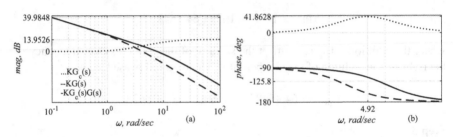

Fig. 8.14 Bode plot of $KG_c(s)G(s)$ and its components: (a) magnitude and (b) phase

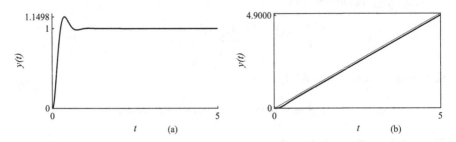

Fig. 8.15 (a) The unit-step response of the lead-compensated system and (b) the ramp response

The Bode plot of the transfer function is shown in Fig. 8.14, along with its components. The corresponding closed-loop transfer function is

$$H(s) = \frac{100.2s + 215}{s^3 + 12.76s^2 + 121.7s + 215}$$

The response is shown in Fig. 8.15. If gain adjustment alone is not sufficient to reduce the overshoot, the zero has to be pushed to the right with the same β. That is ω_{max} has to be increased suitably by trial-and-error.

Example 8.4 Design a lead compensator for the system with the open-loop transfer function

$$G(s) = \frac{1}{s(s + 2)}$$

of a unity feedback system. The specifications are (i) the settling time $t_s < 1.5\%$ and (b) the overshoot is to be limited to 10%.

Solution For this problem, we have to determine appropriate phase margin and closed-loop bandwidth.

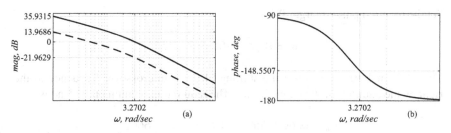

Fig. 8.16 Bode plots of $G(s) = \frac{1}{s(s+2)}$ and $G(s) = \frac{12.5357}{s(s+2)}$: (**a**) magnitude and (**b**) phase

Step 1

Determine ζ and closed-loop bandwidth ω_{BW}. Set the loop-gain K of $G(s)$ to provide the required static error constant and find the corresponding phase margin with this K.

With overshoot 0.1,

$$\zeta = -\frac{\log_e(os)}{\sqrt{\pi^2 + (\log_e(os))^2}} = 0.5912$$

Let us reduce t_s to 1.2 s for a conservative design. Then, the closed-loop bandwidth ω_{BW} is

$$\omega_{\text{BW}} = \left(\frac{4}{\zeta t_s}\right)\sqrt{(1 - 2\zeta^2) + \sqrt{4\zeta^4 - 4\zeta^2 + 2}} = 6.5404$$

rad/sec. Set the crossover frequency one-half of ω_{BW}. $\omega_{\text{Pm}} = 6.5404/2 = 3.2702$ rad/sec. From the Bode plot of $G(s)$, shown in Fig. 8.16a, find the required gain so that the gain is zero at ω_{Pm}. That is, the magnitude plot has to be pushed up by

$$10^{(21.9629/20)} = 12.5357$$

With the gain 12.5357, the transfer function becomes

$$KG(s) = \frac{12.5357}{s(s+2)}$$

The phase at $\omega = 3.2702$ is

$$\left(-\frac{180}{\pi}\right)(\angle(j\omega) + \angle(1 + j\omega/2)) = -148.5507°$$

The phase margin Pm is

$$180° - 148.5507° = 31.4493°$$

Set the required phase margin Pm as

$$Pm = 100\zeta + 10° = 69.1155°$$

The additional 10° is to offset the shifting of the gain crossover frequency to the right by the addition of the compensator. The required maximum phase lead is

$$\theta_{max} = 69.1155° - 31.4493° = 37.6662°$$

Step 2
Compute β and determine the locations of the compensator pole and zero.
   Using θ_{max}, determine β as

$$\beta = \frac{(1 - \sin(\theta_{max}))}{(1 + \sin(\theta_{max}))} = 0.2414$$

Setting $\omega_{max} = \omega_{Pm} = 3.2702$,

$$z\_lead = \omega_{max}\sqrt{\beta} = 1.6068, \qquad p\_lead = z\_lead/\beta = 6.6556$$

The gain K_c of the compensator is

$$K_c = 1/\beta = 4.1422$$

Step 3
Verify that the response of the lead-compensated system meets the specifications, otherwise iterate.
   The transfer function of the lead compensator is

$$G_c(s) = 4.1422\frac{s + 1.6068}{s + 6.6556}$$

The transfer function of the lead-compensated system is

$$G_c(s)G(s) = 4.1422 \left(\frac{s + 1.6068}{s + 6.6556}\right)\left(\frac{12.5357}{s(s + 2)}\right) = \frac{51.93s + 83.43}{s^3 + 8.656s^2 + 13.31s}$$

While this design satisfies the settling time specification, the overshoot is higher than desired. Then, the first choice is to try to solve the problem by gain adjustment. By trial-and-error, the gain of 10.0285 has been found to give the desired response. That is, the transfer function of the lead-compensated system is

$$G_c(s)G(s) = 4.1422 \left(\frac{s + 1.6068}{s + 6.6556}\right)\left(\frac{10.0285}{s(s + 2)}\right) = \frac{41.54s + 66.75}{s^3 + 8.656s^2 + 13.31s}$$

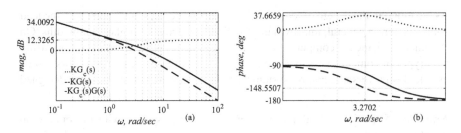

Fig. 8.17 Bode plot of $KG_c(s)G(s)$ along with its components: (**a**) magnitude and (**b**) phase

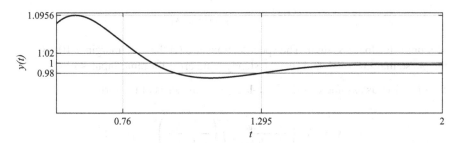

Fig. 8.18 The unit-step response of the lead-compensated system

The Bode plot of the transfer function is shown in Fig. 8.17, along with its components. The corresponding closed-loop transfer function is

$$H(s) = \frac{41.54s + 66.75}{s^3 + 8.656s^2 + 54.85s + 66.75}$$

The response is shown in Fig. 8.18. In case the response does not meet the specifications, the options are to reduce the gain crossover frequency ω_{Pm} or increase ω_{max} or reduce β or a combination of these. Iterate until the response meets the specifications.

8.2.4 Lead–Lag Compensator

This type of compensators is suitable if specifications require the adjustment of bandwidth, phase margin, and steady-state error. It is a combination of that of the lead and lag compensators. The general lead–lag compensator is of the form

$$G_c(s) = K_c \left(\frac{s + z_{lead}}{s + p_{lead}} \right) \left(\frac{s + z_{lag}}{s + p_{lag}} \right), \quad p_{lead} > z_{lead} \quad \text{and} \quad p_{lag} < z_{lag}$$

As we found in the design of lead- and lag-compensated systems separately, the lead compensator, in general, improves the rise time and provides higher bandwidths. On the other hand, the lag compensator improves the steady-state error. Therefore, using both types of compensators in cascade, advantages of both the compensators can be used to design versatile control systems. Either of the compensators can be designed first.

Example 8.5 Design a lead–lag compensator for the system with the open-loop transfer function

$$G(s) = \frac{1}{s(s+1)(s+3)}$$

of a unity feedback system. The specifications are (i) the gain margin $>10\,dB$, (ii) the phase margin $>40°$, and (iii) the steady-state error for ramp input $\leq 12\%$.

Solution Let us assume that the lead–lag compensator is of the form

$$G_c(s) = K_c \left(\frac{s + z_{lead}}{s + \beta z_{lead}} \right) \left(\frac{s + z_{lag}}{s + \frac{z_{lag}}{\beta}} \right), \quad \beta > 1$$

From the required velocity error constant, we get

$$K_v = \frac{1}{0.12} = \lim_{s \to 0} \frac{K}{(s+1)(s+3)} \quad \text{and} \quad K = 25$$

The Bode plot of

$$\frac{25}{s(s+1)(s+3)}$$

is shown in Fig. 8.19. Both the gain and phase margins are negative and the system is unstable. At frequency 1.732 rad/sec, the phase is $-180°$. Let us set the gain crossover frequency at 1.732. At this frequency, sufficient phase margin is to be provided by the compensator. Let us select $z_{lag} = 1.732/10 = 0.1732$, that is, one decade below 1.732. With the specification $\theta_{max} = 40°$,

$$\beta = \frac{(1 + \sin(\theta_{max}))}{(1 - \sin(\theta_{max}))} = 4.5989$$

Let us round 4.5989 to $\beta = 5$, which only increases the phase margin slightly. Then, $p_{lag} = 0.1732/5 = 0.0346$.

Let a be the z_{lead} of the lead compensator. Then, the transfer function of the lead–lag compensator becomes

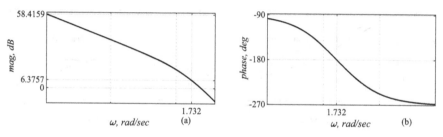

Fig. 8.19 Bode plot of $G(s) = \frac{25}{s(s+1)(s+3)}$: (**a**) magnitude and (**b**) phase

$$G_c(s) = \left(\frac{s+a}{s+(5a)}\right)\left(\frac{s+0.1732}{s+0.0346}\right)$$

At the desired gain crossover frequency 1.732, the gain must be zero dB. However, it is 6.3757 dB, as shown in Fig. 8.20a. Therefore, the gain of the compensator must be -6.3757 dB. Then, using the equation given above, we can find the value of a, as it is the only unknown. The value of a is 0.6953. With this value, the design is complete. However, all the specifications are not satisfied. In particular, steady-state error for ramp input is not acceptable. As the steady-state error is primarily controlled by the lag part of the compensator, we have to redesign it. With $\beta = 5$, the zero is close to the pole and the effectiveness of the pole to improve the steady-state error is reduced. Therefore, $\beta = 6.65$ is found to be suitable by trial-and-error. Now, with the design of the lag part complete and the new β found, the location of the zero of the lead part is found to be 0.5047 by solving for a. Now, the transfer function of the lead–lag compensator is

$$G_c(s) = \left(\frac{s+0.5047}{s+3.3563}\right)\left(\frac{s+0.1732}{s+0.0260}\right)$$

The transfer function of the lead-compensated system is

$$KK_cG_c(s)G(s) = \left(\frac{s+0.5047}{s+3.3563}\right)\left(\frac{s+0.1732}{s+0.0260}\right)\left(\frac{25}{s(s+1)(s+3)}\right)$$

$$= \frac{25s^2 + 16.95s + 2.185}{s^5 + 7.382s^4 + 16.62s^3 + 10.5s^2 + 0.2622s}$$

The corresponding closed-loop transfer function is

$$H(s) = \frac{25s^2 + 16.95s + 2.185}{s^5 + 7.382s^4 + 16.62s^3 + 35.5s^2 + 17.21s + 2.185}$$

The Bode plot of the transfer function is shown in Fig. 8.20, along with its components. The gain and phase margins have been satisfied. The unit-ramp response of

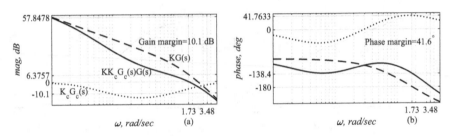

Fig. 8.20 Bode plot of $KK_cG_c(s)G(s)$, $K_cG_c(s)$ and $KG(s)$: (a) magnitude and (b) phase

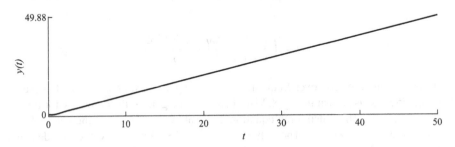

Fig. 8.21 The unit-ramp response of the lead–lag-compensated system

the lead–lag-compensated system is shown in Fig. 8.21 with 12% steady-state error for ramp input.

8.2.5 Proportional–Integral–Derivative Compensator

We have designed PID compensator using root locus. Now, we do the design using Ziegler–Nichols tuning rule, although variations of it are available. The design of PID compensator using this method is applicable, irrespective of the availability of the mathematical model of systems, such as open-loop transfer function. While it may or may not provide optimal control, it proved to be useful in providing satisfactory control and widely used. Furthermore, the PID compensator can be tuned by on-site experiments on the plant. Note that in the purely analytical method, where a mathematical model is required, the uncertainty of proper modeling exists due to nonlinearity and increasing complexity of systems. The PID compensator is of the form

$$G_c(s) = K_p + \frac{K_i}{s} + K_d s = \frac{K_d s^2 + K_p s + K_i}{s}$$

The tuning method provides the initial values of K_p, K_i, and K_d based on the transient response of the system for a step input.

Table 8.2 Ziegler–Nichols tuning rule based on Step Response of system (First method)

Type	K_p	K_i	K_d
P	$\frac{T}{L}$	0	0
PI	$0.9\frac{T}{L}$	$\frac{0.3}{L}$	0
PID	$1.2\frac{T}{L}$	$\frac{1}{2L}$	$0.5L$

The First Method

This method applies to systems with no integrators or dominant complex-conjugate poles. That is, the unit-step response has no overshoot or oscillations. In this method, the unit-step response of the given system is found experimentally or analytically. The response is characterized by two parameters, L and T. These constants are determined as explained in the following example. Then, the system may be approximated by the first-order system with transfer function

$$\frac{Ke^{-Ls}}{Ts + 1}$$

where K is the steady-state value of the response, L is the delay time, and T is the time constant. The parameters of the PID compensator are defined as given in Table 8.2. The transfer function of the PID compensator is

$$G_c(s) = K_p\left(1 + \frac{K_i}{s} + K_d s\right) = 1.2\frac{T}{L}\left(1 + \frac{1}{2Ls} + 0.5Ls\right) = 0.6T\frac{\left(s + \frac{1}{L}\right)^2}{s}$$

The PID compensator has a pole at the origin and double zeros at $s = -1/L$. The pole makes the compensated system Type 1, ensuring zero steady-state error for unit-step input.

Purely derivative or integral plus derivative variations are almost never used.

Example 8.6 Using Ziegler–Nichols tuning rule based on the step response (the first method), design a PID compensator for the system with the open-loop transfer function

$$L(s) = G(s) = \frac{1}{(s + 1)(s + 2)}$$

and the overshoot is to be about 10%.

Solution The determination of L and T is shown in Fig. 8.22. The unit-step response of $G(s)$ is shown by a thick line. The steady-state value is $K = 0.5$. The parameters L and T are determined by drawing a tangent line, shown by a thin line with slope 0.25 (the steepest tangent line), at the infection point of the response. The intercept of this line with the time axis and the horizontal line with $y = K = 0.5$ are at $t = 0.193 = L$ and $t = 2.194$. The value of $T = 2.194 - 0.193 = 2.001$.

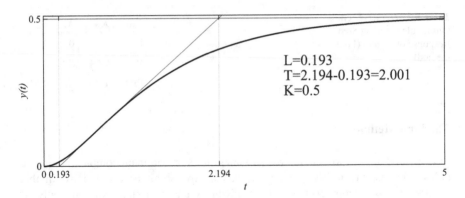

Fig. 8.22 Determining the parameters L and T

A point of inflection is found where the graph of a function changes concavity (curving inwards like the inside of a sphere). A necessary condition that x_0 is an inflection point is that the second derivative of a function $y = f(x)$ is zero; that is, $y''(x_0) = 0$. A sufficient condition is that $y''(x_0 + \epsilon)$ and $y''(x_0 - \epsilon)$, in the neighborhood of x_0, to have opposite signs. Given a sufficiently dense set of uniformly sampled values of the response $y(n)$, the derivative is approximated by a first difference of the samples to get a numerical approximation of $y'(n)$. The accuracy of the approximation depends on the sampling interval. Repeating the operation again on $y'(n)$, we get $y''(n)$. In this set, we look for the value 0. It gives the location of the inflection point. At this location, the value of the slope is found from $y'(n)$. Once a point (t_1, y_1) and the slope m at the point are given, the tangent line is defined by the point-slope form of a straight line as

$$\frac{y - y_1}{t - t_1} = m$$

This is the way, the tangent line shown in the figure is obtained.

With $L = 0.193$ and $T = 2.001$ determined, the transfer function of the PID compensator is

$$G_c(s) = 0.6T\frac{\left(s + \frac{1}{L}\right)^2}{s} = 1.2006\frac{(s + 5.1813)^2}{s}$$

With both $G_c(s)$ and $G(s)$ known, the open-loop transfer function of the PID-compensated system is

$$G_c(s)G(s) = \frac{1.201s^2 + 12.44s + 32.23}{s^3 + 3s^2 + 2s}$$

and the corresponding closed-loop transfer function is

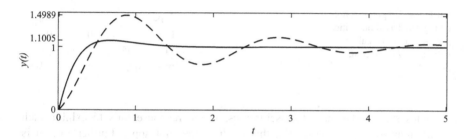

Fig. 8.23 The unit-step response of the PID-compensated system

Table 8.3 A set of solutions with overshoot about 10%

Gain, K	pole, p	y_{max}
4.0000	2.3000	1.1098
4.4000	2.3000	1.1066
4.8000	2.3000	1.1035
5.2000	2.3000	1.1005

$$H(s) = \frac{1.201s^2 + 12.44s + 32.23}{s^3 + 4.201s^2 + 14.44s + 32.23}$$

The unit-step response of the PID-compensated system is shown in Fig. 8.23 by a dashed line. The maximum overshoot is 49.89%, which is quite excessive.

Now, the task is to fine tune this initial design of the compensator so that the specification is met. The PID compensator is of the form

$$G_c(s) = K \frac{(s + p)^2}{s}$$

with K and p as parameters. The rise time, settling time, and overshoot can be part of the specifications. The task is to find a set K and p so that the response meets the specifications. Fine tuning, we get the set of solutions with overshoot about 10%, shown in Table 8.3. The solution with the smallest overshoot can also be found. A MATLAB program to find the set of solutions is available at the book's website. The response with $K = 5.2$ and $p = 2.3$ is also shown in Fig. 8.23 by a solid line. The selection of a particular solution from the table is a trade-off between slow and fast transient responses. Higher gains provide shorter settling time.

Second Method

This method applies to open-loop systems that can be rendered unstable with proportional control. To start with, set $K_i = 0$ and $K_d = 0$. Increase the proportional gain K_p, until the response of the corresponding unity feedback closed-loop system

Table 8.4 Ziegler–Nichols tuning rule based on Critical gain K_{cr} and Critical Frequency F_{cr} (Second method)

Type	K_p	K_i	K_d
P	$0.5K_{cr}$	0	0
PI	$0.45K_{cr}$	$1.2F_{cr}$	0
PID	$0.6K_{cr}$	$2F_{cr}$	$\frac{0.125}{F_{cr}}$

becomes sustained sinusoidal oscillations. If the response fails to exhibit such oscillations for any value of K_p, this method does not apply. Furthermore, it is assumed that the open-loop system is stable. The gain K_p, at this point, is the critical gain K_{cr} and the frequency of oscillation in Hz is the critical frequency F_{cr}. Using these two values, either obtained analytically or experimentally, a table is provided to find the initial values of K_i and K_d. This is an initial design of the compensator, which, eventually, has to be fine tuned by trial-and-error until the design meets the specifications. Ziegler–Nichols tuning rule based on critical gain K_{cr} and critical frequency F_{cr} (second method) is shown in Table 8.4.

$$
G_c(s) = K_p \left(1 + \frac{K_i}{s} + K_d s\right) = 0.6K_{cr}\left(1 + \frac{2F_{cr}}{s} + \frac{0.125}{F_{cr}}s\right)
$$

$$
= \frac{0.075K_{cr}}{F_{cr}} \frac{(s + 4F_{cr})^2}{s}
$$

The PID compensator has a pole at the origin and double zeros at $s = -4F_{cr}$.

Example 8.7 Using Ziegler–Nichols tuning rule based on critical gain K_{cr} and critical frequency F_{cr} (second method), design a PID compensator for the system with the open-loop transfer function with unity feedback

$$
G(s) = \frac{1}{s(s + 1)(s + 3)}
$$

and the overshoot is to be about 10%.

Solution Since $G(s)$ has an integrator, the second method has to be used. Assuming, initially, that K_i and K_d are set to zero, the characteristic equation of the system, with $G_c(s) = K_p$, is

$$
s^3 + 4s^2 + 3s + K_p
$$

Now, the value of K_p, at the point where the roots are purely imaginary, has to be found. The frequency is ω_{cr} rad/sec and $F_{cr} = \omega_{cr}/(2\pi)$ Hz. The gain at that point is K_{cr}. There are at least five ways to find these two parameters. If the mathematical model of the system is not available, they can be determined experimentally. With the unit-step signal as the input, keep increasing the gain K_p from zero to a critical

Table 8.5 The
Routh–Hurwitz array

s^3	1	3
s^2	4	K_p
s^1	$\frac{12-K_p}{4}$	> 0
s^0	K_p	> 0

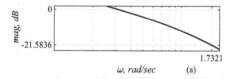

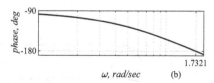

Fig. 8.24 Bode plot of $G(s) = \frac{1}{s(s+1)(s+3)}$: (**a**) magnitude and (**b**) phase

value K_{cr} at which the response exhibits sustained oscillations. The frequency of oscillation is F_{cr}.

The second method is to use the Routh–Hurwitz array, shown in Table 8.5 for this problem (also presented in an earlier chapter). For the system to be stable,

$$\frac{12 - K_p}{4} > 0 \quad \text{and} \quad K_p > 0 \quad \text{or} \quad 0 < K_p < 12$$

At $K_p = K_{cr} = 12$, the characteristic equation is given by

$$s^3 + 4s^2 + 3s + 12$$

and the roots are

$$-4, \quad j1.7321, \quad -j1.7321$$

The frequency is $\omega_{cr} = 1.7321$ rad/sec and $F_{cr} = \omega_{cr}/(2\pi) = 0.2757$ Hz. The gain at that point is K_{cr}. For $K_{cr} < 12$, all the poles are in the left-half of the s-plane.

Even without knowing the value of K_{cr}, we can find ω_{cr}. Substituting $s = j\omega$ in

$$s^3 + 4s^2 + 3s + K_p = 0$$

and confining ourselves only to the imaginary terms, we get

$$-\omega^2 + 3 = 0 \quad \text{or} \quad \omega = \pm\sqrt{3}$$

Iterating $s^3 + 4s^2 + 3s + K_p$ with various values of K_p and finding the roots, we can determine the gain at frequency when the roots include $\pm\sqrt{3}$.

From the root locus or Bode plot of $G(s)$ also, we can find the two required parameter values. The gain margin is 21.5836 dB and gain $K_p = 10^{21.5836/20} = 12$ at the frequency $\omega = \sqrt{3}$ rad/sec, as can be seen from Fig. 8.24. From the root locus,

Fig. 8.25 Root locus of $\dfrac{K}{s(s+1)(s+3)}$

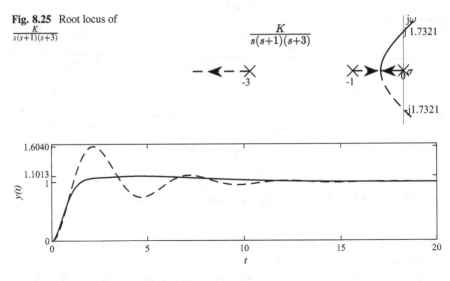

Fig. 8.26 The unit-step response of the PID-compensated system

at frequency $\omega = \sqrt{3}$ rad/sec, the locus crosses the imaginary axis, as can be seen from Fig. 8.25. At that frequency, the gain of the transfer function is computed to be 12.

Once the values of F_{cr} and K_{cr} are known, using Table 8.4, we determine K_p, K_i, and K_d of the PID compensator.

$$\left\{ K_p = 0.6K_{cr} = 7.2, \qquad K_i = 2F_{cr} = 0.5513, \qquad K_d = \frac{0.125}{F_{cr}} = 0.4534 \right\}$$

$$G_c(s) = K_p + \frac{K_i}{s} + K_d s = 7.2 \left(1 + \frac{0.5513}{s} + 0.4534s\right) = 3.2648 \frac{(s+1.1027)^2}{s}$$

With both $G_c(s)$ and $G(s)$ known, the open-loop transfer function of the PID-compensated system is

$$G_c(s)G(s) = \frac{3.265s^2 + 7.2s + 3.97}{s^4 + 4s^3 + 3s^2}$$

and the corresponding closed-loop transfer function is

$$H(s) = \frac{3.265s^2 + 7.2s + 3.97}{s^4 + 4s^3 + 6.265s^2 + 7.2s + 3.97}$$

The unit-step response of the PID-compensated system is shown in Fig. 8.26 by a dashed line. The maximum overshoot is 60.4%, which is quite excessive.

Table 8.6 A set of solutions with overshoot about 10%

Gain, K	Pole, p	y_{max}
0.7	0.6	1.0947
1.0	0.5	1.1072
1.5	0.4	1.0899
1.6	0.4	1.1065
4.2	0.4	1.1196
4.3	0.4	1.1170
4.4	0.4	1.1145
4.5	0.4	1.1121
4.6	0.4	1.1098
4.7	0.4	1.1075
4.8	0.4	1.1054
4.9	0.4	1.1033
5.0	0.4	1.1013

Fine tuning this initial design of the compensator, we get the set of solutions with overshoot about 10%, shown in Table 8.6. The response with $K = 5$ and $p = 0.4$ is also shown in Fig. 8.26 by a solid line.

Example 8.8 Using Ziegler–Nichols tuning rule based on critical gain K_{cr} and critical frequency F_{cr} (the second method), design a PI compensator for the unity feedback system with the open-loop transfer function

$$G(s) = \frac{(s + 0.1)}{s(s + 1)(s + 3)}$$

The specification is that the overshoot is to be as best as possible with the settling time less than 15 s.

Solution The PI compensator is of the form

$$G_c(s) = K \frac{(s + p)}{s}$$

with K and p as parameters. Iteratively searching, we get $K = 10.7$ and $p = 0.4$. The corresponding loop transfer function of the system is

$$G_c(s)G(s) = \frac{10.7s^2 + 5.35s + 0.428}{s^4 + 4s^3 + 3s^2}$$

The unit-step response is shown in Fig. 8.27. The settling time is 14.9510 s and the overshoot is 5.7%.

Example 8.9 Figure 8.28 shows a second-order system with a PID compensator. The reference input $X(s)$ is the unit-step. The disturbance input $D(s)$ is also the

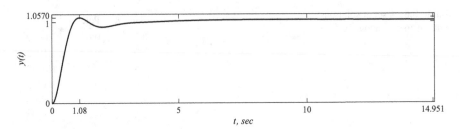

Fig. 8.27 The unit-step response of the PI-compensated system

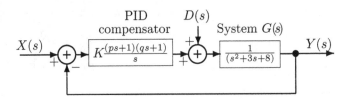

Fig. 8.28 PID compensator for a system with disturbance input

unit-step. The disturbance input should be damped out in about 3 s. Design the PID compensator.

Solution The open-loop transfer function of the system and that of the compensator are

$$G(s) = \frac{1}{(s^2 + 3s + 8)} \quad \text{and} \quad G_c(s) = K\frac{(ps + 1)(qs + 1)}{s}$$

For the disturbance input $D(s)$ alone ($X(s) = 0$), the closed-loop transfer function of the compensated system is

$$Y_d(s) = G(s)(D(s) - Y_d(s)G_c(s)) \quad \text{and} \quad \frac{Y_d(s)}{D(s)} = \frac{1}{\frac{1}{G(s)} + G_c(s)}$$

For this example,

$$\frac{Y_d(s)}{D(s)} = \frac{s}{s(s^2 + 3s + 8) + K(ps + 1)(qs + 1)}$$

$$= \frac{s}{s^3 + (Kpq + 3)s^2 + (Kp + Kq + 8)s + K} \tag{8.1}$$

Since the settling time is 3 s, one reasonable possibility for damping coefficient and natural frequency is $\zeta = 0.4$ and $\omega_n = 3.3333$ rad/sec. Let us choose the far-off real pole at $s = -12$. The desired characteristic equation is

$$(s + 12)(s^2 + 2 \times 0.4 \times 3.3333s + 3.3333^2)$$
$$= (s + 12)(s^2 + 2.6667s + 11.1111)$$
$$= s^3 + 14.6667s^2 + 43.1115s + 133.3332$$

Comparing this equation with Eq. (8.1), we get

$$Kpq + 3 = 14.6667, \quad Kp + Kq + 8 = 43.1115s, \quad K = 133.3332$$

Solving these equations, we get

$$pq = 0.0875, \quad p + q = 0.2633$$

Now,

$$G_c(s) = \frac{K(pqs^2 + (p+q)s + 1)}{s} = \frac{11.6667(s^2 + 3.0095s + 11.4285)}{s}$$

The output to the disturbance alone is

$$Y_d(s) = \frac{s}{s^2 + 14.6667s + 43.1115s + 133.3332} D(s)$$

The closed-loop transfer function for the reference input $X(s)$ is

$$H(s) = \frac{11.6667(s^2 + 3.0095s + 11.4285)}{s^2 + 14.6667s + 43.1115s + 133.3332}$$

The settling time for the reference input is 0.97 s. The output due to disturbance input dies down in about 3 s, as shown in Fig. 8.29.

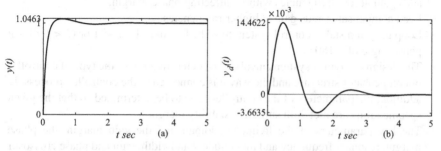

Fig. 8.29 (a) Unit-step input response of $H(s) = \frac{11.6667(s^2+3.0095s+11.4285)}{s^2+14.6667s+43.1115s+133.3332}$ and (b) unit-step disturbance response of $\frac{s}{s^2+14.6667s+43.1115s+133.3332}$

8.3   Summary

- The design of an appropriate compensator is the major task in control system design.
- The frequency response of a system is its response to everlasting sinusoids with infinite frequencies.
- The plot of the magnitude of the complex-valued frequency response versus frequency ω is called the magnitude spectrum. The plot of the phase angle of the complex-valued response versus frequency ω is called the phase spectrum.
- The frequency response of a linear time-invariant system is unique. But, in the Bode and Nyquist plots, the response is plotted in different formats.
- The two plots of the logarithmic magnitude in dB and phase of the frequency response of a system, with logarithmic frequency scales, are known as Bode plots.
- At a corner frequency, the slope of the asymptotic approximation of the frequency response changes or the frequency at which the two magnitude asymptotes intersect.
- The closeness of the $G(j\omega)$ locus to the $(-1, j0)$ point is an indicator of the stability of a system. This requires two measures in the Bode plots. The phase crossover frequency is the frequency at which the phase of $G(j\omega)$ is $-180°$. The gain margin is defined as $-20\log 10|G(j\omega)|$ at the phase crossover frequency. That is, it indicates the additional gain required to bring a stable minimum-phase system to the verge of instability.
- The gain crossover frequency is the frequency at which the gain of $G(j\omega)$ is 0 dB. The phase margin is defined as $180 + \angle G(j\omega)$ at the gain crossover frequency, with $\angle G(j\omega)$ being negative. That is, it indicates the additional phase lag required to bring the system to the verge of instability. For minimum-phase systems, both the measures have to be positive for a stable system.
- The lag compensator is essentially a lowpass filter, suppressing high frequency components. It improves the steady-state error. However, the phase margin is reduced due to the negative phase. This problem is reduced by the zero component of the compensator placed appropriately. That is, lag compensator adds gain at low frequencies without affecting phase margin.
- Lead compensator adds gain at higher frequencies.
- Loop gain of a stable control system must be less than 1 ($G < 1$ or $G < 0$ dB) at phase angle of $-180°$
- The design of control systems involves the determination of the type of controller or compensator structure and the way it is connected to the controlled process. In addition, the parameters of the controller have to be determined so that the given specifications are met with sufficient stability margin.
- The key parameters in the frequency-domain are the gain margin, the phase margin, resonant frequency and magnitude, bandwidth, gain and phase crossover frequencies.
- PID compensators are commonly designed using the Ziegler–Nichols tuning rule.

Exercises

* **8.1** Design a lag compensator for the system with the loop transfer function

$$G(s) = \frac{1}{(s+1)(s+3)}$$

of a unity feedback system. The specifications are (i) steady-state error $e_{ss} <$ 2.5% for step input and (ii) the overshoot is to be limited to 13%.

8.2 Design a lead compensator for the system with the open-loop transfer function

$$G(s) = \frac{1}{s(s+3)}$$

of unity feedback system. The specifications are (i) steady-state error $e_{ss} <$ 15% for ramp input and (b) the overshoot is to be limited to 7%.

* **8.3** Design a lead compensator for the system with the open-loop transfer function

$$G(s) = \frac{1}{s(s+1)}$$

of a unity feedback system. The specifications are (i) the settling time $t_s <$ 1.5% and (b) the overshoot is to be limited to 12%.

8.4 Consider the loop transfer function

$$\frac{K}{(s+1)(s+2)(s+3)}$$

of a unity feedback system. The design specification is that the overshoot is to be limited to about 15%, the steady-state error is to be about 8% with respect to a constant reference, and the settling time is to be about 3 s. Use the lag-lead compensator.

* **8.5** Using Ziegler–Nichols tuning rule based on the step response (the first method), design a PID compensator for the system with the open-loop transfer function

$$L(s) = G(s) = \frac{1}{(s+1)(s+3)}$$

and the overshoot is to be about 12%.

8.6 Using Ziegler–Nichols tuning rule based on critical gain K_{cr} and critical frequency F_{cr} (the second method), design a PID compensator for the system with the open-loop transfer function with unity feedback

$$G(s) = \frac{1}{s(s+1)(s+3)}$$

and the overshoot is to be about 8%.

* **8.7** Using Ziegler–Nichols tuning rule based on critical gain K_{cr} and critical frequency F_{cr} (the second method), design a PI compensator for the unity feedback system with the open-loop transfer function

$$G(s) = \frac{(s+0.1)}{s(s+1)(s+3)}$$

The specification is that the overshoot is to be as best as possible with the settling time less than 20 s.

Chapter 9
Nyquist Plot

A Nyquist plot is a parametric plot of the frequency response, more often used for assessing the stability of a feedback system. It is plotted using polar coordinates, the magnitude and the phase of the frequency response. The magnitude (gain) is the radial coordinate, the phase is the corresponding angular coordinate, and the frequency ω is the parameter. For understanding of the parametric plot, consider the parametric representation of the unit-circle

$$x = \cos(\theta) \quad \text{and} \quad y = \cos(\theta)$$

with θ as the parameter. A point (x, y) is on the plot, if and only if there is a value θ such that the two equations generate that point. The parametric form is an implicit representation of the equation $x^2 + y^2 = 1$. The parametric form is often convenient than the Cartesian form in some applications. The relationships between the two coordinate systems are

$$x = \cos(\theta), \ y = \sin(\theta) \text{ and } x^2 + y^2 = 1, \ \tan(\theta) = \frac{y}{x}, \ (x \neq 0)$$

For easy understanding, we have also provided the corresponding Bode plots, as we are quite familiar with them, presented in the last chapter. Furthermore, the comparison makes it clear that both plots represent the unique frequency response of a system, but in different formats.

As the frequency response, in general, is complex-valued, it can be plotted with real and imaginary parts or with the magnitude and phase versus frequency ω. In the Bode plot, the magnitude in dB versus frequency on a logarithmic scale is plotted in one plot and the phase angle versus frequency is plotted in another plot. In the Nyquist plot, a parametric form is used to represent the same information in a single plot. If a Bode plot is already available, we can take the magnitude and phase values from it and plot the same in a parametric form to get the Nyquist plot. The frequency response is usually plotted for the range $\omega = 0$ to $\omega = \infty$ rather than $\omega = -\infty$

© The Author(s), under exclusive license to Springer Nature Switzerland AG 2022
D. Sundararajan, *Control Systems*, https://doi.org/10.1007/978-3-030-98445-8_9

to $\omega = \infty$, as the frequency response of practical systems is conjugate symmetric. That is, the frequency response is symmetrical about the real axis.

9.1 Nyquist Plot

The closed-loop transfer function of a feedback system is

$$H(s) = \frac{G(s)}{1 + G(s)F(s)}$$

where $G(s)$ is the forward transfer function and $F(s)$ is the feedback transfer function. Now, the stability of the closed-loop transfer function can be determined from the polar plot of $1 + G(s)F(s)$ and by investigating its behavior with respect to the critical point, which is the origin in the s-plane. However, it is easier to get the same information about the stability from the plot of $G(s)F(s)$ by observing its behavior with the **critical point** at $(-1 + j0)$. Then, the **Nyquist plot** is a single plot of the magnitude and phase response of the loop transfer function $G(j\omega)H(j\omega)$ in polar coordinates, as frequency ω varies from $-\infty$ to ∞. From this plot, the stability of the closed-loop system $H(s)$ can be determined. This plot also yields the gain and phase margins, in addition to bandwidth, as in the case of Bode plot. The cutoff frequency is the frequency ω_{BW} at which the closed-loop frequency response is 3 dB below its value at $\omega = 0$. Bandwidth ω_{BW} is the frequency range $0 \le \omega \le \omega_{BW}$. The result is that we get a single plot, rather than two plots as in the Bode plot. In contrast to Bode plot, the Nyquist plot can be applied to systems with delay units without any approximation involved. For such systems, Bode plots yield only approximate stability information. While Bode plot is less general, it proves to be a more useful design tool, as the contributions of each individual factor of the loop transfer function are clearly indicated.

The Nyquist plot is plotted using three steps:

1. Decompose the transfer function $G(s)F(s)$ ($G(s)$ for unity feedback systems) into its real and imaginary parts, after replacing s by $j\omega$.
2. For a range of values of ω, determine the magnitude and phase values.
3. Plot the values in the complex plane and determine gain and phase margins.

9.1.1 Nyquist Plots of Simple Transfer Functions

The Nyquist plots can be derived directly and also from the corresponding Bode plot if it is available. In polar plots, if the phase angle is measured from the positive real axis in the counterclockwise direction, it is positive. If it is measured in the clockwise direction, it is negative. Each point on the Nyquist plot is the tip of a

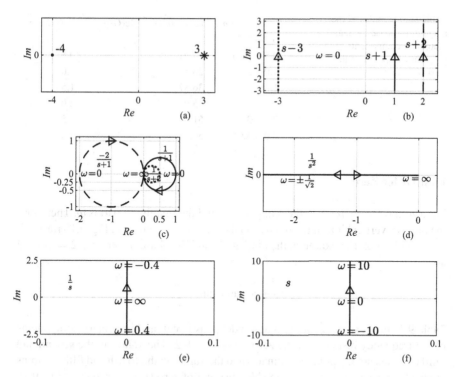

Fig. 9.1 Nyquist plots of simple transfer functions

vector with some magnitude and phase values at a particular frequency ω. Assume that open-loop transfer function $G(s)$ of a unity feedback system is given.

The Constant

The plots of a positive (negative) constant are a point on the positive (negative) side of the real axis, as shown in Fig. 9.1a for $k = -4$ and $k = 3$. If the transfer function evaluates to a real value at any ω, the phase is zero or 180°. Therefore, the plot of a real constant is a point on the real axis. The magnitude and the phase of a constant greater than 1, from Bode plot, are a constant in logarithmic scale and 0°, respectively, for any frequency ω. Therefore, in the Nyquist plot, the representation is the tip of the vector having corresponding magnitude in linear scale and with the same phase angle.

Table 9.1 Samples of the magnitude, phase, real, and imaginary components of the frequency response of the zero $G(s) = s + 2$

ω	$|G(j\omega)|$	$(\angle(G(j\omega))°$	$\text{Re}(G(j\omega))$	$\text{Im}(G(j\omega))$
0	2	0	2	0
0.5	2.06	14.04	2	0.5
1.0	2.24	26.57	2	1.0
1.5	2.5	36.87	2	1.5
2.0	2.83	45.00	2	2.0
2.5	3.2	51.34	2	2.5
3.0	3.61	56.31	2	3.0

First-Order Zero

For a zero $s + a = j\omega + a$, the real part is a and the imaginary part is ω. Therefore, the plot is a vertical line at the point a on the real axis, as shown in Fig. 9.1b for $s - 3$, $s + 1$, and $s + 2$. For example, the magnitude and the phase of zero $s + 2 = j\omega + 2$ are

$$\sqrt{4 + \omega^2} \text{ and } \tan^{-1} \frac{\omega}{2}$$

Table 9.1 shows samples of the magnitude, phase, real, and imaginary components of the frequency response of the zero $G(s) = s + 2$. The values in the second and third columns are the polar coordinates and the values in the fourth and fifth columns are the Cartesian coordinates of the Nyquist plot of zero $(s + 2)$, shown in Fig. 9.1b by a dashed line.

First-Order Pole

For a pole $1/(a + j\omega)$,

$$G(j\omega) = \frac{1}{a + j\omega} = \frac{1}{\sqrt{a^2 + \omega^2}} \angle \left(-\tan^{-1} \left(\frac{\omega}{a} \right) \right)$$

For the specific values $\omega = 0$ and $\omega = a$, we get

$$G(j0) = \frac{1}{a} \angle 0 \quad \text{and} \quad G(ja) = \frac{1}{\sqrt{2a}} \angle - 45°$$

As $\omega \to \infty$, $|G(j\omega)| \to 0$ and $\angle G(j\omega) \to -90°$. The real and imaginary parts of $H(j\omega)$, respectively, are

$$\frac{a}{a^2 + \omega^2} \quad \text{and} \quad \frac{-\omega}{a^2 + \omega^2}$$

Subtracting $1/2a$ from the real part and squaring both the result and adding, we get $(1/2a)^2$. That is,

$$\left(\frac{a}{a^2+\omega^2}-\frac{1}{2a}\right)^2+\left(\frac{\omega}{a^2+\omega^2}\right)^2=\left(\frac{1}{2a}\right)^2$$

This is the equation of a circle with radius $1/(2a)$ and center at $(1/(2a),0)$. Any numerator other than 1 has to be taken into account. The plots for

$$\frac{-2}{1+j\omega}, \quad \frac{1}{1+j\omega}, \quad \frac{1}{2+j\omega}$$

are shown in Fig. 9.1c.

Poles and Zeros at the Origin

Figure 9.1d shows the plot of

$$\frac{1}{s^2}\Big|_{s=j\omega}=-\frac{1}{\omega^2}=\frac{1}{\omega^2}\angle(180°)=\frac{1}{s^2}\Big|_{s=j(-\omega)}$$

which is on the negative real axis and inversely proportional to ω^2. The magnitude of a pole at origin $1/j\omega$ is $1/\omega$ and its phase is constant at $-90°$. Therefore, the plot is along the negative imaginary axis for positive ω and inversely proportional to ω. For a zero, the magnitude and the phase are, respectively, ω and $90°$. Therefore, the plot is along the positive imaginary axis for positive ω and directly proportional to ω. Figure 9.1e and f shows typical examples.

9.2 Stability Analysis from Bode and Nyquist Plots

The transfer function of the closed-loop system, with the forward transfer function $G(s)$ and feedback transfer function $F(s)$, is

$$H(s)=\frac{Y(s)}{X(s)}=\frac{G(s)}{1+G(s)F(s)}$$

While $H(s)$ is the closed-loop transfer function, $G(s)F(s)$ is the loop transfer function. If all the roots of the characteristic equation

$$1+G(s)F(s)=0$$

lie in the left-half of the complex plane, then the system is stable. The poles and zeros of $G(s)F(s)$ may be in the right-half. From the number of poles of $G(s)F(s)$ in the right-half of the complex plane and its Nyquist plot, it is possible to determine both the relative and absolute stability of the closed-loop control system with the transfer function $H(s)$ graphically. This method is directly applicable to systems with delay units also. Furthermore, this stability test can be carried out with experimentally obtained frequency response of the system. The Nyquist stability criterion is based on a theorem from the theory of complex variables. For its use in applications, we can easily understand it, relating it to the stability conditions for the Bode plot. Remember that, for minimum-phase systems, both the Bode and Nyquist plots are the same representations of the frequency response, but in different formats.

Examples
The stability analysis is carried out with both Bode and Nyquist plots for comparison and understanding. Furthermore, the effect of changing the gain and adding poles and zeros to a transfer function can be easily observed in the Bode plots. Figures 9.2, 9.3, 9.4, and 9.5 show Bode plots of some loop transfer functions of unity feedback control systems indicating gain and phase margins.

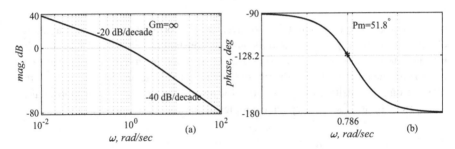

Fig. 9.2 Bode plot of $\frac{1}{(s(s+1))}$ showing gain and phase margins

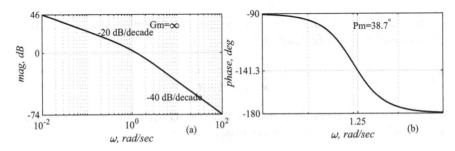

Fig. 9.3 Bode plot of $\frac{2}{(s(s+1))}$ showing gain and phase margins

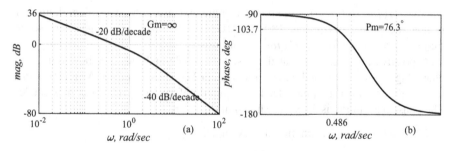

Fig. 9.4 Bode plot of $\frac{1}{(s(s+2))}$ showing gain and phase margins

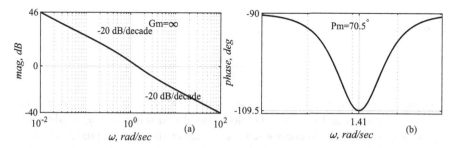

Fig. 9.5 Bode plot of $\frac{s+2}{(s(s+1))}$ showing gain and phase margins

Gain Margin The phase crossover frequency is the frequency ω_p, at which the phase angle of the open-loop transfer function is equal to $-180°$. The gain margin, Gm, is the reciprocal of the magnitude of the frequency response, in decibels, of the system at ω_p.

$$Gm = -20\log_{10}|G(\omega_p)H(\omega_p)|$$

A positive Gm in dB indicates that the system is stable and also how much the gain can be increased before the system becomes unstable. A negative gain margin indicates how much the gain has to be decreased before the system becomes stable. For good performance of the system, the gain has to be high. However, a high gain may make the system unstable. Therefore, the gain has to be set as high as possible with adequate gain margin. The gain margin Gm is ∞ at infinite radians/sec if the phase of $G(\omega)H(\omega)$ is always greater than $-180°$, as shown in Fig. 9.2a. Figure 9.2 shows the Bode plot of $\frac{1}{(s(s+1))}$.

Phase Margin The gain crossover frequency is the frequency ω_g, at which the gain of the open-loop transfer function is equal to 1 or 0 dB. The phase margin, Pm, is defined as

$$Pm = \angle(G(j\omega_g)H(j\omega_g)) - 180°$$

where $\angle(G(j\omega_g)H(j\omega_g))$ is the positive angle in degrees. An equivalent definition is that the **phase margin** is $180° + \angle(G(j\omega_g)H(j\omega_g))$ at the gain crossover frequency, where $\angle(G(j\omega_g)H(j\omega_g))$ is negative.

A positive Pm indicates that the system is stable and indicates how much the phase lag can be increased before the system becomes unstable. A negative Pm indicates how much the phase lead is required before the system becomes stable. The phase margin Pm is 51.8°, as shown in Fig. 9.2b.

Given a loop transfer function, for example,

$$\frac{1}{s(s+1)}$$

First, replace s by $j\omega$ to get

$$\frac{1}{j\omega(j\omega+1)}$$

Compute the magnitude and the phase of the above expression for a range of finite values of frequency ω. Construct two plots, one plot showing the magnitude in dB versus ω in logarithmic frequency scale and another showing the phase versus ω. Always check the computer-generated plots for few values computed manually. Not that the computer has made a mistake. The check is for not entering the transfer function in the program code correctly.

Figure 9.3 shows the Bode plot of $\frac{2}{(s(s+1))}$. The phase margin is less than that in Fig. 9.2b, indicating that the system has become less stable because of the increase in gain. As can be seen in the figures, the increase in the gain does not change the phase response. But the magnitude plot is just pushed up increasing the gain crossover frequency. As the phase lag of the system increases with increasing frequency, the phase margin gets reduced.

Figure 9.4 shows the Bode plot of $\frac{1}{(s(s+2))}$. The phase margin is more than that in Fig. 9.2b, indicating that the system has become more stable because the pole is located further away from the origin.

Figure 9.5 shows the Bode plot of $\frac{(s+2)}{(s(s+1))}$. The phase margin is more than that in Fig. 9.2b, indicating that the system has become more stable because of adding a zero. Although not shown, the system becomes less stable by adding a pole.

9.2.1 Nyquist Stability Criterion

Closed-Loop Stability from the Nyquist Plot

The gain margin is the change in the open-loop gain required to make the closed-loop system unstable. Phase margin is the change in the open-loop phase required to make the closed-loop system unstable.

- Let NP_c be the number of poles (the number of zeros of the characteristic equation) of the closed-loop transfer function in the right-half of s-plane.
- Let NP_l be the number of poles of the loop transfer function in the right-half of s-plane.
- Let N be the number of **encirclements** (to form a circle around the point) of the $(-1,0)$ point by the Nyquist plot. Clockwise encirclements count positive. The number N of clockwise encirclements of the critical point can be found by drawing a vector from that point to the locus, as ω varies in the range $-\infty, \ldots, -1, 0, 1, \ldots, \infty$. The number of rotations the vector makes is N. Counterclockwise encirclements count negative.
- The Nyquist stability criterion is stated as

$$NP_c = NP_l + N$$

For a stable closed-loop system, $NP_c = 0$; that is, $N = -NP_l$. This means that the number of counterclockwise encirclements of the critical point must be NP_l. If $NP_l = 0$, for stability, there must be no encirclement of the critical point. In this case, it is sufficient to ensure that the critical point lies to the left when the plot is traversed for the positive frequency range alone. Otherwise, the closed-loop system is unstable. Now, we know NP_l from the given loop transfer function. We know N from the Nyquist plot. If $NP_l + N \geq 1$, the closed-loop system is unstable. We illustrate this stability criterion through the following examples.

To get the Nyquist plot, plot a single graph with the real part values of the frequency response along the x-axis and the imaginary part values along the y-axis, with ω as a parameter. Figure 9.6 shows the Nyquist plot of the loop transfer function $\frac{1}{(s(s+1))}$. The characteristic equation of the closed-loop transfer function corresponding to the open-loop transfer function $\frac{1}{(s(s+1))}$ is

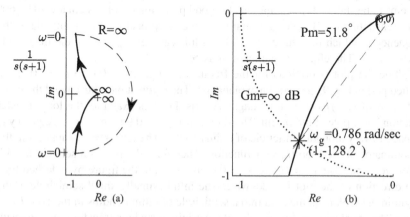

Fig. 9.6 (a) Nyquist plot of $\frac{1}{(s(s+1))}$; (b) part of (a) showing gain and phase margins

$$1 + \frac{1}{s(s+1)} = s^2 + s + 1 = 0$$

The two roots are $-0.5 \pm j0.866$. They are located in the left-half of the complex plane. Therefore, the system is stable. This conclusion, which is absolute stability, can be obtained from the Nyquist plot of $\frac{1}{(s(s+1))}$, shown in Fig. 9.6. The poles of $\frac{1}{(s(s+1))}$ are 0 and -1. That is, the number of poles in the right-half plane is zero, $NP_l = 0$. The Nyquist plot does not encircle the $-1 + j0$ point and the critical point lie to the left when the plot is traversed for the positive frequency range alone. Therefore, $N = 0$ and $NP_c = NP_l + N$, and the number of the poles in the right-half of the complex plane of the corresponding closed-loop transfer function is zero. The system is stable. However, the Nyquist plot does not give the location of the poles. If the degree of the denominator polynomial is greater than that of the numerator, the locus converges clockwise to the origin. At $\omega = \infty$, the locus is tangent to the real axis.

The relative stability can also be found from the Nyquist plot. The magnitude and the phase at the gain cross frequency $\omega_g = 0.786$ rad are, respectively, 1 and $-128.2°$. Therefore, the phase margin Pm is

$$(360° - 128.2°) - 180° = 51.8°$$

as shown in the figure. The phase margin is the angle from the negative real axis to the line (dashed line in the figure) joining the origin and the point where $\omega_g = 0.786$, as shown in Fig. 9.6b. The gain cross over frequency is located at the intersection of the Nyquist plot and the unit-circle, shown by a dotted line in the figure. The gain margin of first- and second-order systems is infinite (since the maximum phase is limited to $180°$) and such systems are always stable. Both the gain and phase margins, as a pair, are a measure of the closeness of the plot to the (-1,0) point and they are design criteria of the system. Both must be positive for minimum-phase systems to be stable. Typical margins for good performance of system are $30° - 60°$ and greater than $6\,dB$. The gain margin is the reciprocal of the magnitude at the frequency at which the phase is $-180°$. In this case, as $\omega_p \to \infty$, the magnitude tends to zero. Therefore, the gain margin is ∞.

While the plot is obvious for the frequency range $\omega = 0+$ to $\omega = 0-$, the dashed part in Fig. 9.6a needs an explanation. The Nyquist plot must avoid the poles and zeros of $G(s)F(s)$ on the imaginary axis. For this example, the loop transfer function has a pole at the origin. Therefore, from $\omega = 0+$ to $\omega = 0-$, frequency ω is made to vary along a semicircle of radius $\epsilon \ll 1$. The term $s = j\omega$ appears in the denominator of the loop transfer function. Therefore, the plot is a semicircle with radius $R = \infty$ from $\omega = 0-$ to $\omega = 0+$, as shown in the figure by a dashed line. The direction of the plot is clockwise. The infinitesimally small semicircle at the origin in the s-plane is mapped into a semicircle of infinite radius in the $G(s)F(s)$-plane. The conclusion is that N poles at the origin in the loop transfer function result in $N/2$ clockwise rotations at infinite radius of the Nyquist plot. For this example,

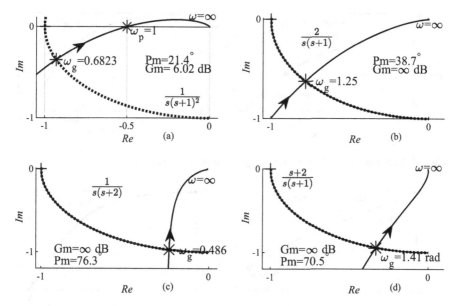

Fig. 9.7 Nyquist plots of (a) $\frac{1}{(s(s+1)^2)}$, (b) $\frac{2}{(s(s+1))}$, (c) $\frac{1}{(s(s+2))}$, and (d) $\frac{(s+2)}{(s(s+1))}$

with one pole at the origin, there is a one-half clockwise rotation in the plot in going from $\omega = 0-$ to $\omega = 0+$.

In Fig. 9.7a, a pole is added and both the gain and phase margins get reduced, making the system less stable. The phase crossover frequency, shown in Fig. 9.7a, is $\omega_p = 1$ rad/sec. At that frequency, the distance between the origin and the phase crossover point is 0.5. Then,

$$Gm = 20 \log 10 \left(\frac{1}{0.5}\right) = 20 \log 10 \, (2) = 6.0206 \text{ dB}$$

Depending on the phase crossover frequency ω_p, we can draw the following conclusions about the gain margin of the system:

1. If the Nyquist plot does not intersect the negative real axis at any finite nonzero frequency, the gain margin Gm is infinite.
2. If the Nyquist plot intersects the negative real axis at the $(-1, j0)$ point, the gain margin Gm is 0 dB.
3. If the Nyquist plot intersects the negative real axis between the origin and $(-1, j0)$ point, the gain margin Gm is positive in dB.
4. If the Nyquist plot intersects the negative real axis to the left of the $(-1, j0)$ point, the gain margin Gm is negative in dB.

The gain crossover frequency, shown in Fig. 9.7a, is $\omega_g = 0.6823$ rad/sec. Then, the phase margin is the angle between the negative real axis and the line that passes

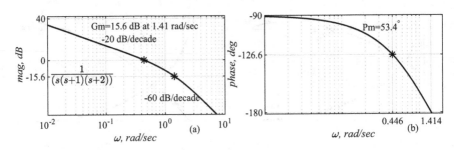

Fig. 9.8 Bode plot of $\frac{1}{(s(s+1)(s+2))}$ showing gain and phase margins

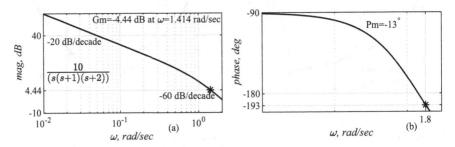

Fig. 9.9 Bode plot of $\frac{10}{(s(s+1)(s+2))}$ showing gain and phase margins

through the gain crossover point and the origin. For this example, the phase margin is 21.4°.

In Fig. 9.7b, the gain is increased and the phase margin gets reduced. In Fig. 9.7c, the pole is located farther from the $j\omega$ axis in the complex plane and the phase margin gets increased. In Fig. 9.7d, a zero is added and the phase margin gets increased. As the gain is increased, the phase crossover frequency moves toward the $-1 + j0$ point and the gain margin gets reduced. That is, the plot gets magnified with the same coordinate system. Therefore, as the gain is further increased, the plot, eventually, encircles the $-1 + j0$ point and the system becomes unstable.

Example 9.1 Figure 9.8 shows the Bode plot of $\frac{1}{(s(s+1)(s+2))}$ showing gain and phase margins. The system is stable with both the gain and phase margins positive. Figure 9.9 shows the Bode plot of $\frac{10}{(s(s+1)(s+2))}$ showing gain and phase margins. The system has become unstable with both the gain and phase margins negative, due to the increase in gain.

Figure 9.10a and b show the Nyquist plots of $\frac{1}{(s(s+1)(s+2))}$ and $\frac{10}{(s(s+1)(s+2))}$, respectively, showing gain and phase margins. The characteristic equation of the closed-loop transfer function corresponding to the open-loop transfer function $\frac{1}{(s(s+1)(s+2))}$ is

$$1 + \frac{1}{s(s+1)(s+2)} = s^3 + 3s^2 + 2s + 1 = 0$$

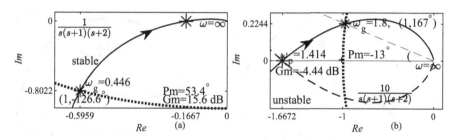

Fig. 9.10 Nyquist plot of (a) $\frac{1}{(s(s+1)(s+2))}$ and (b) $\frac{10}{(s(s+1)(s+2))}$ showing gain and phase margins

The three roots are

$$-2.3247, \quad -0.3376 \pm j0.5623$$

They are all located in the left-half of the complex plane. Therefore, the system is stable. This conclusion, which is absolute stability, can be obtained from the Nyquist plot of $\frac{1}{(s(s+1)(s+2))}$, shown in Fig. 9.10a. The poles of $\frac{1}{(s(s+1)(s+2))}$ are 0, -2, and -1. That is, the number of poles in the right-half plane is zero, $NP_l = 0$. The Nyquist plot does not encircle the $-1 + j0$ point and the critical point lies to the left when the plot is traversed for the positive frequency range alone. Therefore, $N = 0$ and $NP_c = NP_l + N$, the number of the poles in the right-half of the complex plane of the corresponding closed-loop transfer function is zero. The system is stable.

In Fig. 9.10b, the phase crossover frequency occurs at 1.414 rad/sec to the left of the critical point and the gain margin has become negative.

$$Gm = 20 \log 10 \left(\frac{1}{1.6672} \right) = -4.44 \text{ dB}$$

The line between the origin and the gain crossover frequency ω_g is in the second quadrant, and hence, the phase margin is negative. The characteristic equation of the closed-loop transfer function corresponding to the open-loop transfer function $\frac{10}{(s(s+1)(s+2))}$ is

$$1 + \frac{10}{s(s+1)(s+2)} = s^3 + 3s^2 + 2s + 10 = 0$$

The three roots are

$$-3.3089, \quad 0.1545 \pm j1.7316$$

Two of them are located in the right-half of the complex plane. Therefore, the system is unstable. This conclusion, which is absolute stability, can be obtained from the Nyquist plot of $\frac{10}{(s(s+1)(s+2))}$, shown in Fig. 9.10b. The poles of $\frac{10}{(s(s+1)(s+2))}$ are 0,

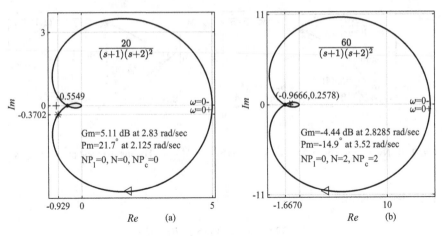

Fig. 9.11 (a) Nyquist plot of $\frac{20}{(s+1)(s+2)^2}$; (b) Nyquist plot of $\frac{60}{(s+1)(s+2)^2}$

-2, and -1. That is, the number of poles in the right-half plane is zero, $NP_l = 0$. The Nyquist plot encircles the $-1 + j0$ point twice, including the one-half rotation due to the pole at origin (not shown in the figure). Therefore, $N = 2$ and $NP_c = NP_l + N$, and the number of the poles in the right-half of the complex plane of the corresponding closed-loop transfer function is two. The system is unstable.

Figure 9.11a shows the Nyquist plot of the open-loop transfer function

$$\frac{20}{(s + 1)(s + 2)^2}$$

The gain crossover point is in the third quadrant, indicating that the phase margin is positive. The phase crossover point occurs to the right of the $(-1, 0)$ point, indicating positive gain margin. The system is stable. Applying the Nyquist criterion for stability also, we come to the same conclusion. That is, the number of open-loop poles NP_l in the right-half plane is zero. The Nyquist plot does not encircle the $-1 + j0$ point and the critical point lies to the left of the plot. Therefore, $N = 0$ and $NP_c = 0 + 0 = 0$. The number of the poles of the corresponding closed-loop transfer function in the right-half is zero.

The characteristic equation of the closed-loop transfer function corresponding to the open-loop transfer function is

$$1 + G(s)F(s) = s^3 + 5s^2 + 8s + 24$$

The roots are

$$\{-4.4187, -0.2906 + j2.3124, -0.2906 - j2.3124\}$$

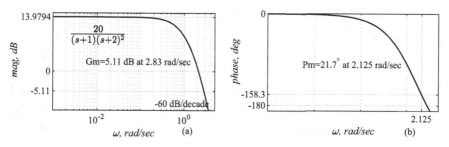

Fig. 9.12 Bode plot of $\frac{20}{(s+1)(s+2)^2}$, showing gain and phase margins

As all the three roots lie in the left-half of the s-plane, the system is stable. The corresponding Bode plot with positive gain and phase margins is shown in Fig. 9.12.

Figure 9.11b shows the Nyquist plot of the open-loop transfer function

$$\frac{60}{(s+1)(s+2)^2}$$

with increased gain. The gain crossover point is in the second quadrant, indicating that the phase margin is negative. The phase crossover point occurs to the left of the $(-1, 0)$ point, indicating negative gain margin. The system is unstable. Applying the Nyquist criterion for stability also, we come to the same conclusion. That is, the number of open-loop poles NP_l in the right-half plane is zero. The Nyquist plot encircles the $-1 + j0$ point two times. Therefore, $N = 2$ and $NP_c = 2 + 0 = 2$. The number of the poles of the corresponding closed-loop transfer function in the right-half of the s-plane is two.

The characteristic equation of the closed-loop transfer function corresponding to the open-loop transfer function is

$$1 + G(s)F(s) = s^3 + 5s^2 + 8s + 64$$

The roots are

$$\{-5.6083, 0.3042 + j3.3644, 0.3042 - j3.3644\}$$

As two of the three roots lie in the right-half of the s-plane, the system is unstable. The corresponding Bode plot with negative gain and phase margins is shown in Fig. 9.13.

For minimum-phase systems, both the Bode and Nyquist plots contain the same information in different formats. The Nyquist plot is of loopy nature. Consider the Nyquist plot shown in Fig. 9.11a. Whenever the phase changes by 180°, the plot has to make one-half of a loop in the polar plot. As the phase increases further, it similarly makes another one-half of a loop, and so on. However, the size of the loops becomes smaller because the magnitude of the frequency response keeps decreasing

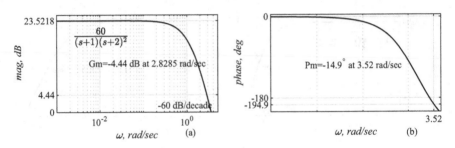

Fig. 9.13 Bode plot of $\frac{60}{(s+1)(s+2)^2}$

with increasing frequencies. The two loops in Fig. 9.11a do not encircle the critical point. The two loops in Fig. 9.11b, however, encircle the critical point two times due to the increase in gain.

9.2.2 Nonminimum-Phase Systems

All the poles and zeros of the loop transfer functions of minimum-phase systems lie in the left-half of the s-plane, excluding the origin. Some poles and/or zeros of nonminimum-phase systems lie in the right-half of the s-plane. With the loop transfer function of nonminimum-phase type, the gain crossover can occur in any quadrant and the definition of phase margin for minimum-phase systems does not apply. The Nyquist criterion for the stability of nonminimum-phase systems is available in the literature, but not that simple as in the case of minimum-phase systems. Of course, we can find the stability by finding the roots of the characteristic equation. For nonminimum-phase systems also, if the $(-1, 0)$ point is enclosed, the system is unstable. If the $(-1, 0)$ point is not enclosed, an additional angle condition has to be satisfied for the system to be stable. These conditions are illustrated through examples. The design of control systems remains the same for both minimum-phase and nonminimum-phase systems.

Example 9.2 Consider the loop transfer function of a unity feedback system

$$G(s) = \frac{10(s + 1)}{(s - 3)(s + 2)}$$

Figure 9.14a shows the Nyquist plot. Although one-half of the plot is sufficient for stability analysis, we have shown the whole plot for completeness. The $(-1, 0)$ point is not enclosed by the plot. The number of poles NP_l of the loop transfer function in the right-half of s-plane, for this example, is 1. An angle θ is defined as

$$\theta = -NP_l 180° = -(1)180° = -180°$$

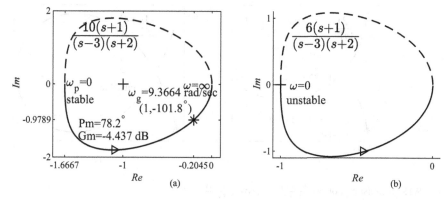

Fig. 9.14 Nyquist plot of (a) $\frac{10(s+1)}{(s-3)(s+2)}$ and (b) $\frac{6(s+1)}{(s-3)(s+2)}$

For the system to be stable, the vector from the $(-1, 0)$ point, shown by the symbol +, to the plot should traverse the angle θ, as ω varies from ∞ to 0. It is true for this example and the system is closed-loop stable, with two roots in the left-half of the s-plane. The corresponding closed-loop transfer function is

$$H(s) = \frac{10s + 10}{s^2 + 9s + 4}$$

The two roots are -8.5311 and -0.4689.

Example 9.3 Consider the loop transfer function of a unity feedback system

$$G(s) = \frac{6(s + 1)}{(s - 3)(s + 2)}$$

The Nyquist plot of the system is shown in Fig. 9.14b and it is unstable. The $(-1 + j0)$ point is on the plot, which indicates that the corresponding closed-loop system has poles on the $j\omega$ axis. Both the gain and phase margins are zero. The corresponding closed-loop transfer function is

$$H(s) = \frac{6s + 6}{s^2 + 5s}$$

The two roots are 0 and -5.

Example 9.4 Consider the loop transfer function of a unity feedback system

$$G(s) = \frac{(-2s + 2)}{(s^2 + 3s)}$$

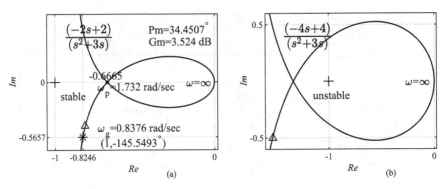

Fig. 9.15 Nyquist plot of (a) $\frac{2(-s+1)}{s(s+3)}$ and (b) $\frac{4(-s+1)}{s(s+3)}$

Figure 9.15a shows the Nyquist plot. The $(-1, 0)$ point is not enclosed by the plot. The number of poles NP_l of the loop transfer function in the right-half of s-plane, for this example, is 0. However, the number of poles NP_{im} of the loop transfer function at the origin of s-plane, for this example, is 1. An angle θ is defined as

$$\theta = -0.5 NP_{im} 180° = -0.5(1)180° = -90°$$

For the system to be stable, the vector from the $(-1, 0)$ point, shown by the symbol $+$, to the plot should traverse the angle θ, as ω varies from ∞ to 0. The plot is a rectangle in the required range. It is true for this example and the system is closed-loop stable, with two roots in the left-half of the s-plane. The corresponding closed-loop transfer function is

$$H(s) = \frac{-2s + 2}{s^2 + s + 2}$$

The two roots are $-0.5 \pm j1.3229$.

Example 9.5 Consider the loop transfer function of a unity-feedback system

$$G(s) = \frac{(-4s + 4)}{(s^2 + 3s)}$$

The system, whose Nyquist plot is shown in Fig. 9.15b, is unstable. The $(-1, 0)$ point is enclosed by the plot. The corresponding closed-loop transfer function is

$$H(s) = \frac{-4s + 4}{s^2 - s + 4}$$

The two roots are $0.5 \pm j1.9365$.

9.2.3 Systems with Delay Units

Figure 9.16 shows Bode plot of $\frac{2}{s+1}$ and $\frac{2e^{-0.4s}}{s+1}$, showing gain and phase margins. For nonminimum-phase systems, the magnitude of the frequency response obtained from both the Bode and Nyquist plots are the same. However, as the phase responses are different, Bode plot of open-loop transfer functions of nonminimum-phase systems should not be used for the stability analysis of the corresponding closed-loop transfer functions. The extra phase reduces both the gain and phase margins compared with the corresponding minimum-phase system.

Figure 9.17a and b show the Nyquist plots of $\frac{2}{s+1}$ and $\frac{2e^{-0.4s}}{s+1}$ with gain and phase margins. The plot is shown only for positive ω in Fig. 9.17a. In Fig. 9.17b, there are several phase crossover frequencies. The gain margin is measured at the highest phase crossover frequency. Similarly, if there are several gain crossover frequencies, the phase margin is measured at the highest gain crossover frequency. The exponential term rotates the transfer function by an angle of $0.4\,\omega$ radians, at each frequency, in the clockwise direction. The magnitude is not affected. The plot of a stable system spirals toward the origin as $\omega \to \infty$. There are an infinite number of intersects on the negative real axis and all of them must occur to the right of $(-1 + j0)$ point for the system to be stable.

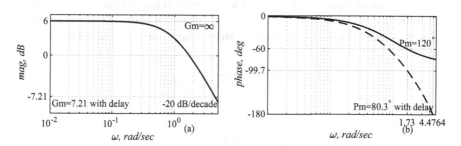

Fig. 9.16 Bode plot of (a) $\frac{2}{s+1}$ and $\frac{2e^{-0.4s}}{s+1}$ showing gain and phase margins: (a) magnitude and (b) phase

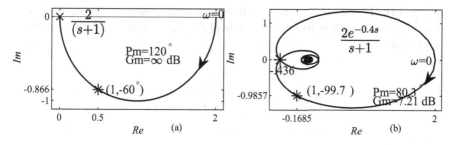

Fig. 9.17 Nyquist plot of (a) $\frac{2}{s+1}$ and (b) $\frac{2e^{-0.4s}}{s+1}$ showing gain and phase margins

9.2.4   Pade Approximation of e^{-Ts}

Pade approximation of the delay function e^{-Ts} is often used to find a rational approximation of transfer functions of systems involving time delays. Let the Nth-order approximant rational function in s be

$$e^{-Ts} \approx \frac{\Sigma_{k=0}^{N} p(k)(Ts)^k}{\Sigma_{k=0}^{N} q(k)(Ts)^k}$$

where T is the time delay in seconds and

$$p(k) = (-1)^k \frac{((2N-k)!N!)}{((2N)!k!(N-k)!)}, \quad q(k) = \frac{((2N-k)!N!)}{((2N)!k!(N-k)!)}, \quad k = 0, 1, \ldots, N$$

For example, applying the formula, we get the Pade approximations for first- and second-order approximants as

$$e^{-Ts} \approx \frac{2 - Ts}{2 + Ts}, \quad \text{and} \quad e^{-Ts} \approx \frac{12 - 6(Ts) + (Ts)^2}{12 + 6(Ts) + (Ts)^2}$$

Let the transfer function be $\frac{2e^{-0.4s}}{s+1}$. The exponential transfer function $e^{-0.4s}$ can be approximated by rational transfer functions using Pade approximation formula. For a first-order approximation, the equivalent rational transfer function is

$$e^{-0.4s} \approx \frac{2 - 0.4s}{2 + 0.4s} = \frac{-s + 5}{s + 5}$$

so that

$$\frac{2e^{-0.4s}}{s+1} \approx 2\frac{-s+5}{(s+5)(s+1)} = \frac{-2s+10}{s^2+6s+5}$$

The system is stable. The Bode plot with positive gain and phase margins is shown in Fig. 9.18. For better accuracy, higher-order approximations can be used. For example, for a second-order approximation of $e^{-0.4s}$, the equivalent rational transfer function is

$$\frac{s^2 - 15s + 75}{s^2 + 15s + 75}$$

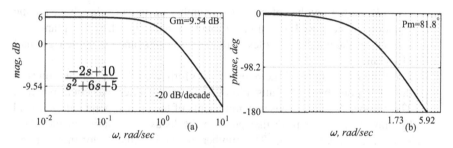

Fig. 9.18 Bode plot of $\frac{2e^{-0.4s}}{s+1}$ (with first-order Pade approximation), showing gain and phase margins: **(a)** magnitude and **(b)** phase

9.3 Summary

- The Nyquist plot is a single plot of the magnitude and phase response of the loop transfer function $G(j\omega)H(j\omega)$ in polar coordinates, as frequency ω varies from $-\infty$ to ∞. Each point on the Nyquist plot is the tip of a vector with some magnitude and phase values at a particular frequency ω.
- The Nyquist plot can be applied to systems with delay units without any approximation involved.
- For minimum-phase systems, both the Bode and Nyquist plots are the same representations of the frequency response, but in different formats.
- The phase crossover frequency is the frequency ω_p, at which the phase angle of the open-loop transfer function is equal to $-180°$. The gain margin, Gm, is the reciprocal of the magnitude of the frequency response, in decibels, of the system at ω_p.

$$\text{Gm} = -20\log_{10}|G(\omega_p)H(\omega_p)|$$

- The gain crossover frequency is the frequency ω_g, at which the gain of the open-loop transfer function is equal to 1 or 0 dB. The phase margin, Pm, is defined as

$$\text{Pm} = \angle(G(j\omega_g)H(j\omega_g)) - 180°$$

where $\angle(G(j\omega_g)H(j\omega_g))$ is the positive angle in degrees.
- Both the gain and phase margins, as a pair, are a measure of the closeness of the plot to the (-1,0) point and they are design criteria of the system. Both must be positive for minimum-phase systems to be stable.
- Nyquist stability criterion: a feedback system is stable if, and only if, the (-1,0) point is on the left of the plot, when the number of poles of the loop transfer function in the right-half of the s-plane is zero.

- - Let NP_c be the number of poles (the number of zeros of the characteristic equation) of the closed-loop transfer function in the right-half of s-plane.
  - Let NP_l be the number of poles of the loop transfer function in the right-half of s-plane.
  - Let N be the number of encirclements (to form a circle around the point) of the (-1,0) point by the Nyquist plot. Clockwise encirclements count positive. The number N of clockwise encirclements of the critical point can be found by drawing a vector from that point to the locus, as ω varies in the range $-\infty, \ldots, -1, 0, 1, \ldots, \infty$. The number of rotations the vector makes is N. Counterclockwise encirclements count negative.
  - The Nyquist stability criterion is stated as

$$NP_c = NP_l + N$$

- All the poles and zeros of the loop transfer functions of minimum-phase systems lie in the left-half of the s-plane, excluding the origin. Some poles and/or zeros of nonminimum-phase systems lie in the right-half of the s-plane.
- The transfer function of systems with delay units can be approximated by rational transfer functions using Pade approximation.

Exercises

9.1 Plot the Bode and Nyquist plots of the loop transfer function $G(s)$ of a unity feedback system. Find the gain and phase margins of the system using both the plots and verify that they are the same. Is the system stable? Find the roots of the corresponding closed-loop transfer function.

9.1.1

$$G(s) = \frac{10}{(s^2 + 3s + 2)}$$

9.1.2

$$G(s) = \frac{10}{s^2 + s + 2}$$

* 9.1.3

$$G(s) = \frac{7(s + 3)}{(s + 1)(s + 2)}$$

9.1.4

$$G(s) = \frac{5}{s^3 + 3s^2 + 2s}$$

* 9.1.5

$$G(s) = \frac{s+1}{s^4 + 5s^3 + 6s^2}$$

9.1.6

$$G(s) = \frac{5s+5}{s^4 + 5s^3 + 6s^2}$$

9.1.7

$$G(s) = \frac{6s+6}{s^4 + 5s^3 + 6s^2}$$

9.1.8

$$G(s) = \frac{s}{s^4 + 5s^3 + 7s^2 + 3s}$$

* 9.1.9

$$G(s) = \frac{1}{s^3 + 2s^2 + s}$$

9.1.10

$$G(s) = \frac{10}{s^3 + 2s^2 + s}$$

9.2 Plot the Nyquist plot of the loop transfer function $G(s)$ of a unity feedback system. Find the gain and phase margins of the system using the plot, if the system stable? Find the roots of the corresponding closed-loop transfer function.

9.2.1

$$G(s) = \frac{10s + 20}{s^2 - 1}$$

* 9.2.2

$$G(s) = \frac{0.3s + 0.6}{s^2 - 1}$$

9.2.3

$$G(s) = \frac{15s + 30}{s^2 - 9}$$

* 9.2.4

$$G(s) = \frac{10s + 20}{s^3 - s}$$

9.3 Plot the Nyquist plot of the loop transfer function $G(s)$ of a unity feedback system. Find the gain and phase margins of the system using the plot if the system is stable?

9.3.1

$$G(s) = \frac{2e^{-0.1s}}{s + 1}$$

* 9.3.2

$$G(s) = \frac{2e^{-0.6s}}{s(s + 1)}$$

9.4 Plot the Bode plot of the first-order Pade approximation of the loop transfer function $G(s)$ of a unity feedback system. Find the gain and phase margins of the system using the plot. Is the system stable? Find the roots of the corresponding closed-loop transfer function.

9.4.1

$$G(s) = \frac{2e^{-0.1s}}{s + 1}$$

* 9.4.2

$$G(s) = \frac{2e^{-0.6s}}{s(s + 1)}$$

Chapter 10
State-Space Analysis of Control Systems

The three system models commonly used in signal and system analysis, differential equation, convolution-integral, and transfer function, are external descriptions of a system. These models do not give a complete description of the dynamical behavior of a system. In some situations, this may be serious shortcoming in the analysis of the system. A system model, called the state-space model, provides a more suitable model that gives the complete description of the dynamical behavior of a system. This is achieved by including the internal states, called the state variables, of the system in addition to the input and output variables in the model of a system. While the model becomes complex, with vector and matrix representation of the variables and fast numerical algorithms of linear algebra, the system can be easily analyzed. For systems of very high order, the state-space analysis is only applicable in practice. The transfer function model is highly suitable for the initial study of the systems but limited to lower-order systems in practice. The advantages of state-space model include application to time-varying and multiple-input and multiple-output systems. The state of a system is the minimum amount of information necessary to completely characterize the behavior of the system at some instant of time. The state variables are the minimum set of variables, which completely describe the state of the system. From the state equations, we can determine any parameter of interest in the system.

A large number of students in engineering disciplines take the circuit theory course in their first year of study followed by the control system course. In the circuit theory course, they study thoroughly loop and nodal analysis of circuits to determine the voltage and currents in various parts of the circuits. Loop and nodal analysis of circuits are basically state-space analysis. Now, we adapt that procedure to the analysis and design of electrical, electro-mechanical, and mechanical feedback systems. Such analysis is also applicable to pneumatic, hydraulic, and thermal feedback systems. While the basic procedure in abstract mathematical form of all types of systems is the same, a lot of details of the individual characteristics of the systems have to be carefully considered in modeling the systems. First, let us recollect the loop and nodal analysis we learnt in the circuit theory course through some examples. In circuit theory, what we call the equilibrium equations are called state equations in state-space analysis. The independent voltage and current variables are called the state variables.

We analyzed circuits in circuit theory with loop and nodal analysis. In these methods, the loop currents and node-to-datum voltages are selected as the independent variables. In state variable analysis, the capacitor voltages and inductor currents are usually selected as independent variables. Both these methods determine all the other currents and voltages in the circuit with some procedural differences. In circuit theory, the concentration is mostly on the steady-state response. The state-space method is popular in control systems, where the total response is of prime interest. The state equations are formulated in a certain form that is particularly suited to use computers in solving them.

While the state variables need not correspond to quantities physically observable in the system, it is a natural choice to choose the output of system components characterized by a first-order differential equation. In electrical circuits, it is a natural choice to choose all the independent capacitor voltages and inductor currents as the state variables. Write the state equations, N for an Nth-order system, using state variables and their first derivatives plus inputs only. That is, we get N simultaneous first-order equations in standard or normal form with the derivative terms alone on the left side of the equation. One possible verification of the state equations is that any possible output of the system at any instant $t > t_0$ can be determined from the state of the system at t_0 with the input known for $t > t_0$.

10.1   The State-Space Model

Example 10.1 Consider the circuit shown in Fig. 10.1. It is a series circuit with a 2 Ω resistor, a 0.5 F capacitor, and the unit-step input voltage source, $u(t)$. Let the current through the circuit be $i(t)$ and the voltages across the resistor and capacitor be, respectively, $v_R(t)$ and $v_C(t)$. Let $v_R(t)$ be the output. The order of the circuit is one as it can be characterized by a first-order differential equation. Let the only state variable be $v_C(t)$, designated as $q(t)$. Let the initial voltage across the capacitor be zero. The general state-space model description, with $x(t)$ input and $y(t)$ output and assuming single input and single output system, is given by

Fig. 10.1 A series circuit
with a resistor and a capacitor

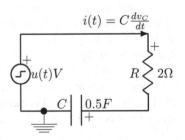

$$i(t) = C\frac{dv_C}{dt}$$

$$\dot{q}(t) = Aq(t) + Bx(t)$$

$$y(t) = Cq(t) + Dx(t)$$

where A is the state matrix, B is the input matrix, C is the output matrix, and D is the transmission matrix. The first equation is always a set of first-order differential equations, called the state equation. This equation involves the state, its first derivative, and the input. The second equation is the output equation. The output is expressed as a linear combination of the state variables and the input.

Applying Kirchhoff's voltage law around the only loop, we get

$$RC\frac{dv_C(t)}{dt} + v_C(t) = u(t) \tag{10.1}$$

$$\dot{q}(t) = -\frac{1}{RC}q(t) + \frac{1}{RC}u(t) \tag{10.2}$$

Using vectors and with $R = 2$ and $C = 0.5$, we get

$$\left[\dot{q}(t)\right] = \left[-1\right]\left[q(t)\right] + \left[1\right]u(t)$$

$$v_R(t) = \left[-1\right]\left[q(t)\right] + \left[1\right]u(t)$$

The state-space model becomes

$$A = \left[-\tfrac{1}{RC}\right], \quad B = \left[\tfrac{1}{RC}\right], \quad C = \left[-1\right], \quad D = 1$$

Figure 10.2 shows the block diagram representation of the state-space model of the first-order system, with single input and single output. Now, the output can be computed either in the time domain or in the frequency domain using the state-space model.

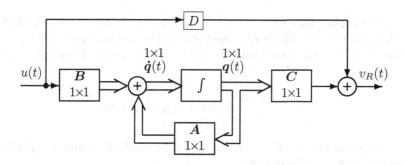

Fig. 10.2 Block diagram representation of the state-space model of the first-order system, with single input and single output

10.2   Frequency-Domain Solution of the State Equation

The Laplace transform of a vector function, such as $q(t)$, is defined to be the vector function $Q(s)$, where the elements are the transforms of the corresponding elements of $q(t)$. Let the initial state vector be $q(0^-)$. Using the time-differentiation property of the Laplace transform, we get the Laplace transform of the state equation as

$$s Q(s) - q(0^-) = A Q(s) + B X(s)$$

Since $I Q(s) = Q(s)$, where I is the identity matrix of the same size as the matrix A, and collecting the terms involving $Q(s)$, we get

$$(s I - A) Q(s) = q(0^-) + B X(s)$$

The inclusion of the identity matrix is necessary to combine the terms involving $Q(s)$. Premultiplying both sides by $(s I - A)^{-1}$, which is the inverse of $(s I - A)$, we get

$$Q(s) = \overbrace{(s I - A)^{-1} q(0^-)}^{q_{zi}(s)} + \overbrace{(s I - A)^{-1} B X(s)}^{q_{zs}(s)}$$

The inverse Laplace transforms of the first and the second expressions on the right-hand side are, respectively, the zero-input and zero-state components of the state vector $q(t)$. Taking the Laplace transform of the output equation, we get

$$Y(s) = C Q(s) + D X(s)$$

Substituting for $Q(s)$, we get

$$Y(s) = \overbrace{C(s I - A)^{-1} q(0^-)}^{y_{zi}(s)} + \overbrace{(C(s I - A)^{-1} B + D) X(s)}^{y_{zs}(s)}$$

The inverse Laplace transforms of the first and the second expressions on the right-hand side are, respectively, the zero-input and zero-state components of the system response $y(t)$. With the system initial conditions zero, the transfer function is given by

$$H(s) = \frac{Y(s)}{X(s)} = (C(s I - A)^{-1} B + D)$$

By setting proper values to C and D, we can find the transfer function corresponding to various variables in the system.

Example 10.2 Solve the problem of Example 10.1 using the frequency-domain method.

Solution

$$(sI - A) = s[1] - [-\tfrac{1}{RC}] = [s + \tfrac{1}{RC}]$$

The zero-state component of the state vector is

$$q_{zs}(s) = (sI - A)^{-1}BX(s) = \frac{1}{RC}\frac{1}{s(s + \tfrac{1}{RC})} = \frac{1}{s} - \frac{1}{s + \tfrac{1}{RC}}$$

Taking the inverse Laplace transform, we get

$$q(t) = v_C(t) = (1 - e^{-\frac{t}{RC}})u(t)$$

$$i(t) = C\frac{dv_C(t)}{dt} = \frac{1}{R}e^{-\frac{t}{RC}}u(t)$$

With the voltage across the resistor $v_R(t)$ as the output $y(t)$, we get

$$y(t) = [-1]q(t) + u(t) = e^{-\frac{t}{RC}}u(t)$$

With $RC = 1$,

$$i(t) = 0.5e^{-t}u(t), \quad v_R(t) = e^{-t}u(t), \quad v_C(t) = (1 - e^{-t})u(t)$$

Initial Condition

Let the initial voltage across the capacitor be $v_C(0) = 3V$. Find the zero-input response of the circuit.

$$Y_{zi}(s) = C(sI - A)^{-1}q(0^-)$$

With $C = 1$, $q(0^-) = 3$, and

$$(sI - A)^{-1} = \frac{1}{s + \tfrac{1}{RC}}$$

we get

$$Y_{zi}(s) = V_R(s) = -\frac{3}{s + \tfrac{1}{RC}}$$

Taking the inverse Laplace transform, we get

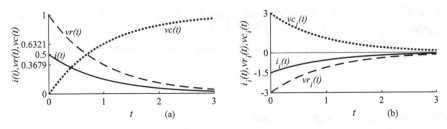

Fig. 10.3 The response of the series RC circuit: (**a**) to the unit-step voltage input alone and (**b**) to the initial voltage 3 V across the capacitor alone

$$v_R(t) = -3(e^{-\frac{t}{RC}})u(t)$$

Dividing $v_R(t)$ by R, we get

$$i(t) = -\frac{3}{R}e^{-\frac{t}{RC}}u(t)$$

Assuming that clockwise direction is positive for the current flow, the current due the initial condition is negative.

$$v_C(t) = -v_R(t) = 3(e^{-\frac{t}{RC}})u(t)$$

The response of the series RC circuit to the unit-step voltage input alone and to the initial voltage 3 V across the capacitor alone is shown, respectively, in Fig. 10.3a and b. The time constant of the circuit is $RC = 1$. In one time constant, the value of $v_R(t)$ reduces to 0.3679 of its initial value. The time constant is useful for comparison of different responses. The value of $v_C(t)$ raises to $1 - 0.3679 = 0.6321$ of its final value. For practical purposes, the duration of 4 time constants is considered as the duration of the transient response. In that period, the response increases to 98% of its final value for a growing exponential and the response reduces to below 2% for a decaying exponential. The total response of the circuit is the sum of the responses to the input and the initial condition.

Example 10.3 Find the voltage $v_1(t)$ across the inductor in the circuit shown in Fig. 10.4, in both time and frequency domains, using the state-space analysis.

Solution The suggestive procedure for state-space analysis is as follows: (i) choose all independent inductor currents and capacitor voltages as state variables, (ii) select a set of currents through the loops and express the state variables and their first derivatives as a linear combination of the loop currents, and (iii) using the loop equations, eliminate all other variables except the state variables and their first derivatives.

$$R_1 = 2\Omega, \quad L_2 = 0.3H, \quad C_3 = 0.5F, \quad v_2 = u(t)V$$

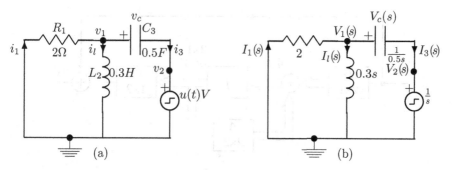

Fig. 10.4 A circuit with unit-step input voltage, in the time domain on the left and in the frequency domain on the right

The capacitor voltage v_C and the inductor current i_l are the natural choices as the state variables. Applying Kirchhoff's voltage law around the loop involving resistor and capacitor, we get

$$v_R = i_1 R_1 = -v_c - u(t) \quad \text{or} \quad i_1 = -0.5v_c - 0.5u(t)$$

$$C_3 \frac{dv_c}{dt} = i_1 - i_l = -0.5v_c - 0.5u(t) - i_l$$

The voltage across the inductor must be equal to the sum of $v_c + u(t)$.

$$L_2 \frac{di_l}{dt} = v_c + u(t)$$

Simplifying and writing in standard form, we get the two state equations as

$$\frac{dv_c}{dt} = -v_c - 2i_l - u(t)$$

$$\frac{di_l}{dt} = \frac{1}{0.3}v_c + \frac{1}{0.3}u(t)$$

Let $q_1(t) = v_c(t)$ and $q_2(t) = i_l(t)$ be the two state variables. Using matrices, we get the state-space model of the circuit as

$$\dot{q}(t) = \begin{bmatrix} \dot{q}_1(t) \\ \dot{q}_2(t) \end{bmatrix} = \begin{bmatrix} -1 & -2 \\ \frac{1}{0.3} & 0 \end{bmatrix} \begin{bmatrix} q_1(t) \\ q_2(t) \end{bmatrix} + \begin{bmatrix} -1 \\ \frac{1}{0.3} \end{bmatrix} [u(t)]$$

$$y(t) = \begin{bmatrix} 1 & 0 \end{bmatrix} \begin{bmatrix} q_1(t) \\ q_2(t) \end{bmatrix} + u(t)$$

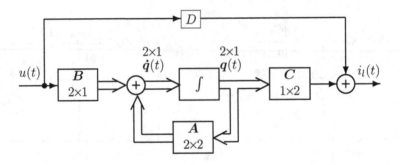

Fig. 10.5 Block diagram representation of the state-space model of the second-order system, with single input and single output

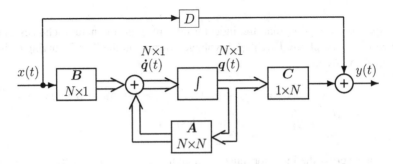

Fig. 10.6 Block diagram representation of the state-space model of an Nth-order system, with single input and single output

assuming that v_l is the output. Therefore,

$$A = \begin{bmatrix} -1 & -2 \\ \frac{1}{0.3} & 0 \end{bmatrix}, \quad B = \begin{bmatrix} -1 \\ \frac{1}{0.3} \end{bmatrix}, C = \begin{bmatrix} 1 & 0 \end{bmatrix}, \quad D = 1$$

Figure 10.5 shows the block diagram representation of the state-space model of the second-order system, with single input and single output. In general, the block diagram representation of the state-space model of an Nth-order system, with single input and single output, is shown in Fig. 10.6. Parallel lines terminating with an arrowhead indicate that the signal is a vector quantity.

Example 10.4 Solve the problem of Example 10.3 using the frequency-domain method.

Solution

$$(sI - A) = s \begin{bmatrix} 1 & 0 \\ 0 & 1 \end{bmatrix} - \begin{bmatrix} -1 & -2 \\ \frac{1}{0.3} & 0 \end{bmatrix} = \begin{bmatrix} s+1 & 2 \\ -\frac{1}{0.3} & s \end{bmatrix}$$

The inverse, A^{-1}, of a 2×2 matrix

$$A = \begin{bmatrix} a & b \\ c & d \end{bmatrix} \text{ is defined as } A^{-1} = \frac{1}{ad-bc} \begin{bmatrix} d & -b \\ -c & a \end{bmatrix}, \quad ad-bc \neq 0$$

$$(sI-A)^{-1} = \frac{1}{s^2+s+\frac{20}{3}} \begin{bmatrix} s & -2 \\ \frac{1}{0.3} & s+1 \end{bmatrix} = \begin{bmatrix} \frac{s}{s^2+s+\frac{20}{3}} & -\frac{2}{s^2+s+\frac{20}{3}} \\ \frac{\frac{10}{3}}{s^2+s+\frac{20}{3}} & \frac{s+1}{s^2+s+\frac{20}{3}} \end{bmatrix}$$

In the frequency domain, a check of the inverse, using the initial value theorem of the Laplace transform, is $\lim_{s\to\infty} s(sI-A)^{-1} = I$.

The zero-state component of the state vector is

$$q_{zs}(s) = (sI-A)^{-1}BX(s) = \begin{bmatrix} \frac{s}{s^2+s+\frac{20}{3}} & -\frac{2}{s^2+s+\frac{20}{3}} \\ \frac{\frac{10}{3}}{s^2+s+\frac{20}{3}} & \frac{s+1}{s^2+s+\frac{20}{3}} \end{bmatrix} \begin{bmatrix} -1 \\ \frac{1}{0.3} \end{bmatrix} \frac{1}{s}$$

$$= \begin{bmatrix} -\frac{(s+\frac{20}{3})}{s(s^2+s+\frac{20}{3})} \\ \frac{\frac{10}{3}s}{s(s^2+s+\frac{20}{3})} \end{bmatrix} = \begin{bmatrix} -\frac{1}{s} + \frac{0.5+j0.0987}{s+0.5-j2.5331} + \frac{0.5-j0.0987}{s+0.5+j2.5331} \\ \frac{0.5}{s} + \frac{-0.25+j0.0493}{s+0.5-j2.5331} + \frac{-0.25-j0.0493}{s+0.5+j2.5331} \end{bmatrix}$$

Taking the inverse Laplace transform, we get

$$q_{zs}(t) = \begin{bmatrix} (-1+1.0193e^{-0.5t}\cos(2.5331t+0.1949))u(t) \\ (1.3159e^{-0.5t}\cos(2.5331t-0.5\pi))u(t) \end{bmatrix}$$

The first and second components of $q_{zs}(t)$ are, respectively, v_c and i_l due to the input alone.   ∎

Let the initial values of the state variables are

$$q_1(0) = 1 \quad \text{and} \quad q_2(0) = 0.5$$

The initial state vector is

$$q(0^-) = \begin{bmatrix} 1 \\ 0.5 \end{bmatrix}$$

$$q_{zi}(s) = (sI-A)^{-1}q(0^-) = \begin{bmatrix} \frac{s}{s^2+s+\frac{20}{3}} & -\frac{2}{s^2+s+\frac{20}{3}} \\ \frac{\frac{10}{3}}{s^2+s+\frac{20}{3}} & \frac{s+1}{s^2+s+\frac{20}{3}} \end{bmatrix} \begin{bmatrix} 1 \\ 0.5 \end{bmatrix} = \begin{bmatrix} \frac{s-1}{s^2+s+\frac{20}{3}} \\ \frac{0.5s+\frac{11.5}{3}}{s^2+s+\frac{20}{3}} \end{bmatrix}$$

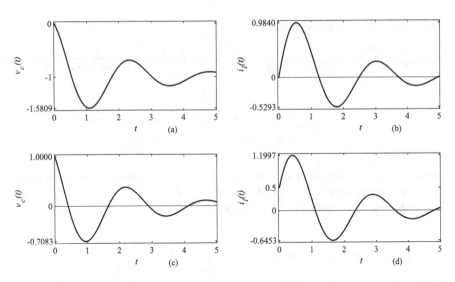

Fig. 10.7 (a) Zero-state response v_c, (b) zero-state response i_l, (c) zero-input response v_c, and (d) zero-input response i_l

Taking the inverse Laplace transform, we get

$$\boldsymbol{q}_{zi}(t) = \begin{bmatrix} (1.1622e^{-0.5t}\cos(2.5331t + 0.5346)u(t) \\ (1.5004e^{-0.5t}\cos(2.5331t - 1.2310))u(t) \end{bmatrix}$$

The first and second components of $\boldsymbol{q}_{zi}(t)$ are, respectively, v_c and i_l due to the initial conditions alone. The zero-state response v_c, zero-state response i_l, zero-input response v_c, and zero-input response i_l are shown in Fig. 10.7a–d. From the equation for zero-input response of current i_l and its derivative, we get, as $t \to 0$, $\dot{y}(0^-) = 10/3$ and $y(0^-) = 1/2$. These values are usually given as initial conditions. Then, the initial state vector has to be derived from the given initial output conditions using the state and output equations. From the state equation, we get

$$\dot{q}_1(0^-) = -q_1(0^-) - 2q_2(0^-)$$

$$\dot{q}_2(0^-) = \frac{10}{3}q_1(0^-)$$

Note that the input $x(t)$ is zero at $t = 0^-$. From the output equation, we get

$$0\dot{q}_1(0^-) + \dot{q}_2(0^-) = \frac{10}{3}$$

$$0q_1(0^-) + q_2(0^-) = 0.5$$

There are four unknowns and four equations. Solving these equations, we get the initial state vector as

$$q_1(0^-) = 1, \qquad q_2(0^-) = 0.5$$

as given initially.

Characteristic Roots (Eigenvalues) of a Matrix

The characteristic polynomial of a system is given by the determinant of the matrix $(s\boldsymbol{I} - \boldsymbol{A})$, where $\boldsymbol{I}$ is the identity matrix of the same size as $\boldsymbol{A}$. While we can write down the characteristic polynomial from the differential equation, we just show how it can be found using the matrix $\boldsymbol{A}$. For this example,

$$(s\boldsymbol{I} - \boldsymbol{A}) = s\begin{bmatrix} 1 & 0 \\ 0 & 1 \end{bmatrix} - \begin{bmatrix} -1 & -2 \\ 10/3 & 0 \end{bmatrix} = \begin{bmatrix} s+1 & 2 \\ -10/3 & s \end{bmatrix}$$

The characteristic polynomial of the system, given by the determinant of this matrix

$$\begin{vmatrix} s+1 & 2 \\ -10/3 & s \end{vmatrix} \qquad \text{is} \qquad s^2 + s + \frac{20}{3}$$

With each of the infinite different realizations of a system, we get the $\boldsymbol{A}$ matrix with different values. However, as the system is the same, its characteristic polynomial, given by the determinant of $(s\boldsymbol{I} - \boldsymbol{A})$, will be the same for any valid $\boldsymbol{A}$. The characteristic roots or eigenvalues, which are the roots of this polynomial, are $\lambda_1 = -0.5 + j2.5331$ and $\lambda_2 = -0.5 - j2.5331$. These values are the poles of the corresponding transfer function. The zero-input response, of an Nth-order system with N distinct eigenvalues, is of the form

$$y_{zi}(t) = c_1 e^{\lambda_1 t} + c_2 e^{\lambda_2 t} + \cdots + c_N e^{\lambda_N t}$$

10.3   Time-Domain Solution of the State Equation

Let us find the solution of the state equation in the time domain. For this purpose, we need the exponential of a matrix $e^{\boldsymbol{A}t}$ and its derivative. Similar to the infinite series defining an exponential of a scalar,

$$e^{\boldsymbol{A}t} = \boldsymbol{I} + \boldsymbol{A}t + \boldsymbol{A}^2 \frac{t^2}{2!} + \boldsymbol{A}^3 \frac{t^3}{3!} + \cdots$$

Note that $e^0 = \boldsymbol{I}$ and $e^{\boldsymbol{A}t} e^{-\boldsymbol{A}t} = \boldsymbol{I}$. This series is absolutely and uniformly convergent for all values of t. Therefore, it can be differentiated or integrated term by term.

$$\frac{d(e^{At})}{dt} = A + A^2 t + A^3 \frac{t^2}{2!} + A^4 \frac{t^3}{3!} + \cdots = Ae^{At} = e^{At} A$$

By premultiplying both sides of state equation by e^{-At}, we get

$$e^{-At} \dot{q}(t) = e^{-At} Aq(t) + e^{-At} Bx(t)$$

By shifting the first term on the right side to the left, we get

$$e^{-At} \dot{q}(t) - e^{-At} Aq(t) = e^{-At} Bx(t)$$

Since

$$\frac{d(e^{-At} q(t))}{dt} = e^{-At} \dot{q}(t) - e^{-At} Aq(t)$$

we can write the previous equation as

$$\frac{d(e^{-At} q(t))}{dt} = e^{-At} Bx(t)$$

Integrating both sides of this equation from 0^- to t, we get

$$e^{-At} q(t)|_{0^-}^{t} = \int_{0^-}^{t} e^{-A\tau} Bx(\tau) d\tau$$

Applying the limit and then premultiplying both sides by e^{At}, we get

$$q(t) = \overbrace{e^{At} q(0^-)}^{q_{zi}(t)} + \overbrace{\int_{0^-}^{t} e^{A(t-\tau)} Bx(\tau) d\tau}^{q_{zs}(t)}$$

The first and the second expressions on the right-hand side are, respectively, the zero-input and zero-state components of the state vector $q(t)$. Note that the part of the expression

$$\int_{0^-}^{t} e^{A(t-\tau)} Bx(\tau) d\tau$$

is the convolution of the matrices e^{At} and $Bx(t)$, $e^{At} * Bx(t)$. The convolution of matrices is the same as the multiplication of two matrices, except that the product of two elements is replaced by their convolution. If the initial state vector values are given at $t = t_0^-$, rather than at $t = 0^-$, the state equation is modified as

$$q(t) = e^{A(t-t_0)} q(t_0^-) + \int_{t_0^-}^{t} e^{A(t-\tau)} Bx(\tau)d\tau$$

The matrix e^{At} is called the state-transition matrix or the fundamental matrix of the system. The inverse Laplace transform of $((sI - A)^{-1})$ is e^{At}.

Once we know the state vector, we get the output of the system using the output equation as

$$y(t) = C \left(e^{At} q(0^-) + \int_{0^-}^{t} e^{A(t-\tau)} Bx(\tau)d\tau \right) + Dx(t)$$

$$= \overbrace{Ce^{At}q(0^-)}^{y_{zi}(t)} + \overbrace{C \int_{0^-}^{t} e^{A(t-\tau)} Bx(\tau)d\tau + Dx(t)}^{y_{zs}(t)}$$

The first expression on the right-hand side is the zero-input component of the system response $y(t)$ and the other two expressions yield the zero-state component. The zero-input response of the system depends solely on the state-transition matrix e^{At}. This matrix, for an Nth-order system, is evaluated, using Cayley–Hamilton theorem, as

$$e^{At} = c_0 I + c_1 A + c_2 A^2 + \cdots + c_{N-1} A^{(N-1)}$$

where

$$\begin{bmatrix} c_0 \\ c_1 \\ \cdots \\ c_{N-1} \end{bmatrix} = \begin{bmatrix} 1 & \lambda_1 & \lambda_1^2 & \cdots & \lambda_1^{N-1} \\ 1 & \lambda_2 & \lambda_2^2 & \cdots & \lambda_2^{N-1} \\ & & \cdots & & \\ 1 & \lambda_N & \lambda_N^2 & \cdots & \lambda_N^{N-1} \end{bmatrix}^{-1} \begin{bmatrix} e^{\lambda_1 t} \\ e^{\lambda_2 t} \\ \cdots \\ e^{\lambda_N t} \end{bmatrix}$$

and $\lambda_1, \lambda_2, \ldots, \lambda_N$ are the N distinct characteristic roots of A. For a root λ_r repeated m times, the first row corresponding to that root will remain the same as for a distinct root and the $m - 1$ successive rows will be successive derivatives of the first row with respect to λ_r. For example, with the first root of a fourth-order system repeating two times, we get

$$\begin{bmatrix} c_0 \\ c_1 \\ c_2 \\ c_3 \end{bmatrix} = \begin{bmatrix} 1 & \lambda_1 & \lambda_1^2 & \lambda_1^3 \\ 0 & 1 & 2\lambda_1 & 3\lambda_1^2 \\ 1 & \lambda_2 & \lambda_2^2 & \lambda_2^3 \\ 1 & \lambda_3 & \lambda_3^2 & \lambda_3^3 \end{bmatrix}^{-1} \begin{bmatrix} e^{\lambda_1 t} \\ te^{\lambda_1 t} \\ e^{\lambda_2 t} \\ e^{\lambda_3 t} \end{bmatrix}$$

Example 10.5 Solve the problem of Example 10.3 using the time-domain method.

Solution

$$A = \begin{bmatrix} -1 & -2 \\ \frac{10}{3} & 0 \end{bmatrix}, \quad B = \begin{bmatrix} -1 \\ \frac{10}{3} \end{bmatrix}, \quad C = \begin{bmatrix} 1 & 0 \end{bmatrix}, \quad D = 1$$

$$\lambda_1 = -0.5 + j2.5331 \quad \text{and} \quad \lambda_2 = -0.5 - j2.5331$$

The transition matrix is given by

$$e^{At} = c_0 I + c_1 A$$

where

$$\begin{bmatrix} c_0 \\ c_1 \end{bmatrix} = \begin{bmatrix} 1 & \lambda_1 \\ 1 & \lambda_2 \end{bmatrix}^{-1} \begin{bmatrix} e^{\lambda_1 t} \\ e^{\lambda_2 t} \end{bmatrix} = \begin{bmatrix} 0.5 - j0.0987 & 0.5 + j0.0987 \\ -j0.1974 & j0.1974 \end{bmatrix} \begin{bmatrix} e^{(-0.5+j2.5331)t} \\ e^{(-0.5-j2.5331)t} \end{bmatrix}$$

$$= \begin{bmatrix} 1.0193e^{-0.5t} \cos(2.5331t - 0.1949) \\ 0.3948e^{-0.5t} \cos(2.5331t - 0.5\pi) \end{bmatrix}$$

$$e^{At} = c_0 \begin{bmatrix} 1 & 0 \\ 0 & 1 \end{bmatrix} + c_1 \begin{bmatrix} -1 & -2 \\ \frac{10}{3} & 0 \end{bmatrix} = \begin{bmatrix} c_0 - c_1 & -2c_1 \\ \frac{10}{3}c_1 & c_0 \end{bmatrix}$$

$$= \begin{bmatrix} 1.0193e^{-0.5t} \cos(2.5331t + 0.1949) & -0.7896e^{-0.5t} \cos(2.5331t - 0.5\pi) \\ 1.3160e^{-0.5t} \cos(2.5331t - 0.5\pi) & 1.0193e^{-0.5t} \cos(2.5331t - 0.1949) \end{bmatrix}$$

Since $q(t) = e^{At}q(0)$, with $t = 0$, we get $q(0) = e^{A0}q(0)$; that is, $I = e^{A0}$. This result, which can be used to check the state-transition matrix, is also obvious from the infinite series for e^{At}.

The state vector $q_{zi}(t)$ can be computed as follows:

$$q_{zi}(t)$$

$$= \begin{bmatrix} 1.0193e^{-0.5t} \cos(2.5331t + 0.1949) & -0.7896e^{-0.5t} \cos(2.5331t - 0.5\pi) \\ 1.3160e^{-0.5t} \cos(2.5331t - 0.5\pi) & 1.0193e^{-0.5t} \cos(2.5331t - 0.1949) \end{bmatrix} \begin{bmatrix} 1 \\ 0.5 \end{bmatrix}$$

$$= \begin{bmatrix} (1.1622e^{-0.5t} \cos(2.5331t + 0.5346))u(t) \\ (1.5004e^{-0.5t} \cos(2.5331t - 1.2310))u(t) \end{bmatrix}$$

as obtained by the frequency-domain method.

Convolution of two matrices is similar to the multiplication of two matrices with the multiplication of the elements replaced by the convolution of the elements. Convolution of $x(t)$ with the unit-step signal is the running integral of $x(t)$.

$$x(t) * u(t) = \int_{-\infty}^{t} x(\tau)d\tau$$

The state vector $q_{zs}(t)$ can be computed as follows:

$q_{zs}(t)$

$$= \begin{bmatrix} 1.0193e^{-0.5t}\cos(2.5331t + 0.1949) & -0.7896e^{-0.5t}\cos(2.5331t - 0.5\pi) \\ 1.3160e^{-0.5t}\cos(2.5331t - 0.5\pi) & 1.0193e^{-0.5t}\cos(2.5331t - 0.1949) \end{bmatrix}$$

$$* \left(\begin{bmatrix} -1 \\ \frac{10}{3} \end{bmatrix} u(t) \right)$$

$$= \begin{bmatrix} \int_0^t (-1.0193e^{-0.5\tau}\cos(2.5331\tau + 0.1949) - \frac{10}{3}(0.7896)e^{-0.5\tau}\cos(2.5331\tau - 0.5\pi))d\tau \\ \int_0^t (-1.3160e^{-0.5\tau}\cos(2.5331\tau - 0.5\pi) + \frac{10}{3}(1.0193)e^{-0.5\tau}\cos(2.5331\tau - 0.1949))d\tau \end{bmatrix}$$

$$= \begin{bmatrix} -2.6320\int_0^t (e^{-0.5\tau}\cos(2.5331\tau - 1.1811))d\tau \\ 3.3977\int_0^t (e^{-0.5\tau}\cos(2.5331\tau + 0.1949))d\tau \end{bmatrix}$$

$$= \begin{bmatrix} (-1 + 1.0193e^{-0.5t}\cos(2.5331t + 0.1949))u(t) \\ (1.3159e^{-0.5t}\cos(2.5331t - 0.5\pi))u(t) \end{bmatrix}$$

The capacitor voltage converges to -1, as shown in Fig. 10.7a, and also from the corresponding closed-form expression for the voltage as $t \to \infty$. The total response of the system is the sum of the zero-input and zero-state components of the response.

The complex exponential is widely used in the applications of science and engineering. Apart from its role as basis function in major transforms, one of its uses is to simplify the evaluation of integrals involving trigonometric functions. The integrals are first converted to complex exponential form, integrated and the real or imaginary parts of the result are taken, as required. The integration of

$$-2.6320 \int_0^t (e^{-0.5\tau}\cos(2.5331\tau - 1.1811))d\tau$$

is first converted to complex form as

$$2.6320 \int_0^t e^{j\pi} e^{-j1.1811} e^{(-0.5+j2.5331)\tau} d\tau$$

Integrating, we get

$$\frac{2.6320e^{j\pi}e^{-j1.1811}}{(-0.5+j2.5331)} e^{(-0.5+j2.5331)t} \Big|_0^t$$

Taking the real part of the integral, we get

$$(-1 + 1.0193e^{-0.5t}\cos(2.5331t + 0.1949))u(t)$$

Computation of the State-Transition Matrix e^{At}

One method to compute the state-transition matrix is to use the Cayley–Hamilton theorem, as presented in the solution of this problem. Of course, the inverse Laplace transform of $((sI - A)^{-1})$ in the frequency-domain solution is also e^{At}. Another method is to diagonalize the A matrix, as presented later.

10.4   Commonly Used Realizations of Systems

Provided there is no pole-zero cancellation, the transfer function of a system derived from the state-space representation and from system analysis will be the same. While the transfer function is unique, it can be realized in several ways. For example, a 10 Ω resistor can be realized by several ways. A series connection of two 5 Ω resistors or a series connection of ten 1 Ω resistors results in 10 Ω, and so on. The circuit input–output relationship will be the same with any of these realizations.

Controllable Canonical Form

Canonical form realization of systems is the realization with a minimum number of delay units. That number is the same as the order of the system. Consider the state-space model, shown in Fig. 10.8, of a second-order continuous system, characterized by the differential equation

$$\ddot{y}(t) + a_1\dot{y}(t) + a_2 y(t) = b_0\ddot{x}(t) + b_1\dot{x}(t) + b_2 x(t)$$

A dot over a variable indicates its first derivative and two dots indicate its second derivative. For example, $\dot{y}(t) = \frac{dy(t)}{dt}$ and $\ddot{y}(t) = \frac{d^2 y(t)}{dt^2}$. In Fig. 10.8, the input is $x(t)$ and the output is $y(t)$. The two internal variables (called the state variables) of the system are $q_1(t)$ and $q_2(t)$. State variables are a minimal set of variables (N

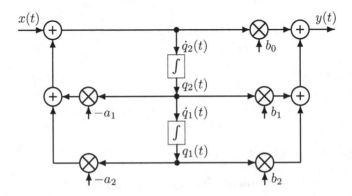

Fig. 10.8 A state-space model of the controllable canonical form of a second-order continuous system

for an Nth-order system) of a system so that a knowledge of the values of these variables (the state of the system) at $t = t_0$ and those of the input for $t \geq t_0$ will enable the determination of the values of the state variables for $t > t_0$ and the output for $t \geq t_0$. An infinite number of different sets, each of N state variables, are possible for a particular Nth-order system.

From Fig. 10.8, we can write down the following state equations defining the state variables $q_1(t)$ and $q_2(t)$:

$$\dot{q}_1(t) = q_2(t)$$
$$\dot{q}_2(t) = -a_1 q_2(t) - a_2 q_1(t) + x(t)$$

The first derivative of each state variable is expressed in terms of all the state variables and the input. No derivatives of either the state variables or the input are permitted, on the right-hand side, to have the equation in a standard form. A second-order differential equation characterizing the system, shown in Fig. 10.8, has been decomposed into a set of two simultaneous first-order differential equations. Selecting state variables as the output of the integrators is a natural choice, since an integrator is characterized by a first-order differential equation. With that choice, we can write down a state equation at the input of each integrator. However, the state variables need not correspond to quantities those are physically observable in a system. In the state-space model of a system, in general, an Nth-order differential equation characterizing a system is decomposed into a set of N simultaneous first-order differential equations of a standard form. With a set of N simultaneous differential equations, we can solve for N unknowns. These are the N internal variables, called the state variables, of the system. The output is expressed as a linear combination of the state variables and the input. The concepts of impulse response, convolution, and transform analysis are all equally applicable to the state-space model. The difference is that, as the system is modeled using matrix and vector quantities, the system analysis involves matrix and vector quantities. One of the advantages of the state-space model is the easier extension to multiple inputs and outputs. For simplicity, we describe systems with single input and single output only. The output $y(t)$ of the system, shown in Fig. 10.8, is given by

$$y(t) = -b_0 a_1 q_2(t) - b_0 a_2 q_1(t)$$
$$+ b_2 q_1(t) + b_1 q_2(t) + b_0 x(t)$$

The output equation is an algebraic (not a differential) equation. We can write the state and output equations, using vectors and matrices, as

$$\begin{bmatrix} \dot{q}_1(t) \\ \dot{q}_2(t) \end{bmatrix} = \begin{bmatrix} 0 & 1 \\ -a_2 & -a_1 \end{bmatrix} \begin{bmatrix} q_1(t) \\ q_2(t) \end{bmatrix} + \begin{bmatrix} 0 \\ 1 \end{bmatrix} x(t)$$

$$y(t) = \begin{bmatrix} b_2 - b_0 a_2 & b_1 - b_0 a_1 \end{bmatrix} \begin{bmatrix} q_1(t) \\ q_2(t) \end{bmatrix} + b_0 x(t)$$

Let us define the state vector $q(t)$ as

$$q(t) = \begin{bmatrix} q_1(t) \\ q_2(t) \end{bmatrix}$$

Then, with

$$A = \begin{bmatrix} 0 & 1 \\ -a_2 & -a_1 \end{bmatrix}, \quad B = \begin{bmatrix} 0 \\ 1 \end{bmatrix},$$

$$C = \begin{bmatrix} b_2 - b_0 a_2 & b_1 - b_0 a_1 \end{bmatrix}, \quad D = b_0$$

The general state-space model description for continuous systems is given as

$$\dot{q}(t) = Aq(t) + Bx(t)$$

$$y(t) = Cq(t) + Dx(t)$$

Let us derive this form from the second-order transfer function

$$H(s) = \frac{Y(s)}{X(s)} = \frac{b_0 s^2 + b_1 s + b_2}{s^2 + a_1 s + a_2}$$

The transfer function can be expressed as a cascade of two transfer functions as

$$H(s) = \left(\frac{1}{s^2 + a_1 s + a_2} \right) (b_0 s^2 + b_1 s + b_2)$$

Let the output of the first transfer function be $Q(s)$. Then,

$$Q(s) = \left(\frac{1}{s^2 + a_1 s + a_2} \right) X(s) \quad \text{and} \quad Y(s) = (b_0 s^2 + b_1 s + b_2) Q(s)$$

The corresponding expression for the first transfer function, in differential equation form, is

$$x(t) = \ddot{q}(t) + a_1 \dot{q}(t) + a_2 q(t) \quad \text{or} \quad \ddot{q}(t) = x(t) - a_1 \dot{q}(t) - a_2 q(t)$$

The realization of this equation is shown in the left side of Fig. 10.8. The necessary variables are available at the integrators to form the output $y(t)$ from $q(t)$, as shown in the right side of Fig. 10.8.

The Nth-order transfer function is of the form

$$H(s) = \frac{Y(s)}{X(s)} = \frac{b_0 s^N + b_1 s^{N-1} + \cdots + b_{N-1} s + b_N}{s^N + a_1 s^{N-1} + \cdots + a_{N-1} s + a_N}$$

By generalizing the result for a second-order system, we get the state-space representation of a N-th order system as

$$
\begin{bmatrix} \dot{q}_1 \\ \dot{q}_2 \\ \vdots \\ \dot{q}_{N-1} \\ \dot{q}_N \end{bmatrix} = \begin{bmatrix} 0 & 1 & 0 \cdots & 0 \\ 0 & 0 & 1 \cdots & 0 \\ \vdots & \vdots & \vdots & \vdots \\ 0 & 0 & 0 \cdots & 1 \\ -a_N & -a_{N-1} & -a_{N-2} \cdots & -a_1 \end{bmatrix} \begin{bmatrix} q_1 \\ q_2 \\ \vdots \\ q_{N-1} \\ q_N \end{bmatrix} + \begin{bmatrix} 0 \\ 0 \\ \vdots \\ 0 \\ 1 \end{bmatrix} x
$$

$$
y = \begin{bmatrix} b_N - a_N b_0 & b_{N-1} - a_{N-1} b_0 & \cdots & b_1 - a_1 b_0 \end{bmatrix} \begin{bmatrix} q_1 \\ q_2 \\ \vdots \\ q_N \end{bmatrix} + b_0 x
$$

Each of the denominator coefficients of the transfer function appears negated in reverse order in the bottom row of the state matrix A, which is an important feature of this form. Input vector B consists of all zeros, except for a 1 in its last row. This form is extremely useful in designing control systems by the pole-placement approach.

Example 10.6 With voltage across the capacitor as the output, the transfer function of the system in Example 10.4 is

$$
H(s) = \frac{-s - \frac{20}{3}}{s^2 + s + \frac{20}{3}}
$$

Find the state-space model corresponding to $H(s)$ in controllable canonical form. Find the transfer function from this model and verify that it is the same as the given one.

Solution The state-space model is

$$
A = \begin{bmatrix} 0 & 1 \\ -a_2 & -a_1 \end{bmatrix} = \begin{bmatrix} 0 & 1 \\ -\frac{20}{3} & -1 \end{bmatrix}, \quad B = \begin{bmatrix} 0 \\ 1 \end{bmatrix},
$$

$$
C = \begin{bmatrix} b_2 - b_0 a_2 & b_1 - b_0 a_1 \end{bmatrix} = \begin{bmatrix} -\frac{20}{3} & -1 \end{bmatrix}, \quad D = b_0 = 0
$$

With the system initial conditions zero, the transfer function is given by

$$
H(s) = \frac{Y(s)}{X(s)} = (C(sI - A)^{-1}B + D)
$$

$$
(sI - A) = s \begin{bmatrix} 1 & 0 \\ 0 & 1 \end{bmatrix} - \begin{bmatrix} 0 & 1 \\ -\frac{20}{3} & -1 \end{bmatrix} = \begin{bmatrix} s & -1 \\ \frac{20}{3} & s + 1 \end{bmatrix}
$$

$$(s\boldsymbol{I} - \boldsymbol{A})^{-1} = \frac{1}{s^2 + s + \frac{20}{3}}\begin{bmatrix} s+1 & 1 \\ -\frac{20}{3} & s \end{bmatrix}$$

$$H(s) = \begin{bmatrix} -\frac{20}{3} & -1 \end{bmatrix}\frac{1}{s^2 + s + \frac{20}{3}}\begin{bmatrix} s+1 & 1 \\ -\frac{20}{3} & s \end{bmatrix}\begin{bmatrix} 0 \\ 1 \end{bmatrix}$$

$$= \frac{-s - \frac{20}{3}}{s^2 + s + \frac{20}{3}}$$

Observable Canonical Form

There is a dual realization that can be derived by using the transpose operation of a matrix. This realization is called observable canonical form and is characterized by the matrices defined, in terms of those of controllable canonical form, as

$$\overline{\boldsymbol{A}} = \boldsymbol{A}^T, \overline{\boldsymbol{B}} = \boldsymbol{C}^T, \overline{\boldsymbol{C}} = \boldsymbol{B}^T, \overline{\boldsymbol{D}} = D$$

Example 10.7 With voltage across the capacitor as the output, the transfer function of the system in Example 10.4 is

$$H(s) = \frac{-s - \frac{20}{3}}{s^2 + s + \frac{20}{3}}$$

Find the state-space model corresponding to $H(s)$ in observable canonical form. Find the transfer function from this model and verify that it is the same as the given one.

Solution The state-space model is

$$\boldsymbol{A} = \begin{bmatrix} 0 & -a_2 \\ 1 & -a_1 \end{bmatrix} = \begin{bmatrix} 0 & -\frac{20}{3} \\ 1 & -1 \end{bmatrix}, \quad \boldsymbol{B} = \begin{bmatrix} b_2 - b_0a_2 \\ b_1 - b_0a_1 \end{bmatrix} = \begin{bmatrix} -\frac{20}{3} \\ -1 \end{bmatrix},$$

$$\boldsymbol{C} = \begin{bmatrix} 0 & 1 \end{bmatrix}, \quad D = b_0 = 0$$

$$(s\boldsymbol{I} - \boldsymbol{A}) = s\begin{bmatrix} 1 & 0 \\ 0 & 1 \end{bmatrix} - \begin{bmatrix} 0 & -\frac{20}{3} \\ 1 & -1 \end{bmatrix} = \begin{bmatrix} s & \frac{20}{3} \\ -1 & s+1 \end{bmatrix}$$

$$(s\boldsymbol{I} - \boldsymbol{A})^{-1} = \frac{1}{s^2 + s + \frac{20}{3}}\begin{bmatrix} s+1 & -\frac{20}{3} \\ 1 & s \end{bmatrix}$$

$$H(s) = \begin{bmatrix} 0 & 1 \end{bmatrix}\frac{1}{s^2 + s + \frac{20}{3}}\begin{bmatrix} s+1 & -\frac{20}{3} \\ 1 & s \end{bmatrix}\begin{bmatrix} -\frac{20}{3} \\ -1 \end{bmatrix}$$

$$= \frac{-s - \frac{20}{3}}{s^2 + s + \frac{20}{3}}$$

Diagonal Canonical Form

Assuming the roots of the denominator polynomial are distinct, we get by partial-fraction expansion

$$H(s) = \frac{b_0 s^N + b_1 s^{N-1} + \cdots + b_{N-1} s + b_N}{s^N + a_1 s^{N-1} + \cdots + a_{N-1} s + a_N}$$

$$= b_0 + \frac{k_1}{s + p_1} + \frac{k_2}{s + p_2} + \cdots + \frac{k_N}{s + p_N}$$

The diagonal canonical form is given by

$$\begin{bmatrix} \dot{q}_1 \\ \dot{q}_2 \\ \vdots \\ \dot{q}_N \end{bmatrix} = \begin{bmatrix} -p_1 & & & 0 \\ & -p_2 & & \\ & & \ddots & \\ 0 & & & -p_N \end{bmatrix} \begin{bmatrix} q_1 \\ q_2 \\ \vdots \\ q_N \end{bmatrix} + \begin{bmatrix} 1 \\ 1 \\ \vdots \\ 1 \end{bmatrix} x$$

$$y = \begin{bmatrix} k_1 & k_2 & \cdots & k_N \end{bmatrix} \begin{bmatrix} q_1 \\ q_2 \\ \vdots \\ q_N \end{bmatrix} + b_0 x$$

The poles (eigenvalues) form the diagonal elements of the diagonal matrix A. Matrix C consists of the partial-fraction coefficients. All the entries of matrix B are 1.

Example 10.8

$$H(s) = \frac{s + 3}{s^2 + 3s + 2}$$

Find the state-space model corresponding to $H(s)$ in diagonal canonical form. Find the transfer function from this model and verify that it is the same as the given one.

Solution The roots of the denominator polynomial of the transfer function are -1 and -2.

$$H(s) = \frac{s + 3}{s^2 + 3s + 2} = \frac{-1}{s + 2} + \frac{2}{s + 1}$$

The state-space model is

$$A = \begin{bmatrix} -p_1 & 0 \\ 0 & -p_2 \end{bmatrix} = \begin{bmatrix} -1 & 0 \\ 0 & -2 \end{bmatrix}, \quad B = \begin{bmatrix} 1 \\ 1 \end{bmatrix},$$

$$C = \begin{bmatrix} 2 & -1 \end{bmatrix}, \quad D = b_0 = 0$$

$$(sI - A) = s \begin{bmatrix} 1 & 0 \\ 0 & 1 \end{bmatrix} - \begin{bmatrix} -1 & 0 \\ 0 & -2 \end{bmatrix} = \begin{bmatrix} s+1 & 0 \\ 0 & s+2 \end{bmatrix}$$

$$(sI - A)^{-1} = \frac{1}{s^2 + 3s + 2} \begin{bmatrix} s+2 & 0 \\ 0 & s+1 \end{bmatrix}$$

$$H(s) = [2 \ -1] \frac{1}{s^2 + 3s + 2} \begin{bmatrix} s+2 & 0 \\ 0 & s+1 \end{bmatrix} \begin{bmatrix} 1 \\ 1 \end{bmatrix}$$

$$= \frac{s+3}{s^2 + 3s + 2}$$

Jordan Canonical Form

A modified form of the diagonal canonical form is Jordan canonical form, which is applicable when all the eigenvalues are not distinct. Let the first two eigenvalues (poles) be the same. Then, we get by partial-fraction expansion

$$H(s) = \frac{b_0 s^N + b_1 s^{N-1} + \cdots + b_{N-1} s + b_N}{s^N + a_1 s^{N-1} + \cdots + a_{N-1} s + a_N}$$

$$= b_0 + \frac{k_1}{(s + p_1)^2} + \frac{k_2}{s + p_1} + \frac{k_3}{s + p_3} + \cdots + \frac{k_N}{s + p_N}$$

The Jordan canonical form is given by

$$\begin{bmatrix} \dot{q}_1 \\ \dot{q}_2 \\ \vdots \\ \dot{q}_N \end{bmatrix} = \begin{bmatrix} -p_1 & 1 & & 0 \\ 0 & -p_1 & & \\ & & \ddots & \\ 0 & & & -p_N \end{bmatrix} \begin{bmatrix} q_1 \\ q_2 \\ \vdots \\ q_N \end{bmatrix} + \begin{bmatrix} 0 \\ 1 \\ \vdots \\ 1 \end{bmatrix} x$$

$$y = [k_1 \ k_2 \ \cdots \ k_N] \begin{bmatrix} q_1 \\ q_2 \\ \vdots \\ q_N \end{bmatrix} + b_0 x$$

The poles (eigenvalues) form the diagonal elements of the diagonal matrix A, with a 1 entry over the repeated poles, except the first one. All the entries of matrix B are 1, except that there is only 1 entry corresponding to the last of the repeated poles with the rest zeros.

Example 10.9 The transfer function of a system is

$$H(s) = \frac{s+3}{s^2 + 4s + 4}$$

Find the state-space model corresponding to $H(s)$ in Jordan canonical form. Find the transfer function from this model and verify that it is the same as the given one.

Solution The roots of the denominator polynomial of the transfer function are -2 and -2.

$$H(s) = \frac{s+3}{s^2+4s+4} = \frac{1}{(s+2)^2} + \frac{1}{s+2}$$

The state-space model is

$$A = \begin{bmatrix} -p_1 & 1 \\ 0 & -p_1 \end{bmatrix} = \begin{bmatrix} -2 & 1 \\ 0 & -2 \end{bmatrix}, \quad B = \begin{bmatrix} 0 \\ 1 \end{bmatrix},$$

$$C = \begin{bmatrix} 1 & 1 \end{bmatrix}, \quad D = b_0 = 0$$

$$(sI - A) = s\begin{bmatrix} 1 & 0 \\ 0 & 1 \end{bmatrix} - \begin{bmatrix} -2 & 1 \\ 0 & -2 \end{bmatrix} = \begin{bmatrix} s+2 & -1 \\ 0 & s+2 \end{bmatrix}$$

$$(sI - A)^{-1} = \frac{1}{s^2+4s+4}\begin{bmatrix} s+2 & 1 \\ 0 & s+2 \end{bmatrix}$$

$$H(s) = \begin{bmatrix} 1 & 1 \end{bmatrix}\frac{1}{s^2+4s+4}\begin{bmatrix} s+2 & 1 \\ 0 & s+2 \end{bmatrix}\begin{bmatrix} 0 \\ 1 \end{bmatrix} + 0 = \frac{s+3}{s^2+4s+4}$$

All canonical forms can be converted from one to another. When diagonal form is not possible, Jordan form may be used.

Example 10.10 The transfer function of a system is

$$H(s) = \frac{s^2+s+2}{s^2+4s+4}$$

Find the state-space model corresponding to $H(s)$ in Jordan canonical form. Find the transfer function from this model and verify that it is the same as the given one.

Solution The roots of the denominator polynomial of the transfer function are -2 and -2.

$$H(s) = \frac{s^2+s+2}{s^2+4s+4} = 1 + \frac{-3s-2}{(s+2)^2} = 1 + \frac{4}{(s+2)^2} + \frac{-3}{s+2}$$

$$= 1 + \frac{1}{(s+2)}\left(-3 + \frac{4}{s+2}\right)$$

The state-space model is

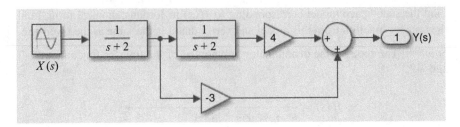

Fig. 10.9 Generation of the output $y(t)$ from input $x(t)$ for the repeated poles only. The direct output is not shown

$$A = \begin{bmatrix} 0 & 1 \\ -4 & -4 \end{bmatrix}, \quad B = \begin{bmatrix} 0 \\ 1 \end{bmatrix},$$

$$C = \begin{bmatrix} 4 & -3 \end{bmatrix}, \quad D = b_0 = 1$$

$$(sI - A) = s\begin{bmatrix} 1 & 0 \\ 0 & 1 \end{bmatrix} - \begin{bmatrix} 0 & 1 \\ -4 & -4 \end{bmatrix} = \begin{bmatrix} s & -1 \\ 4 & s+4 \end{bmatrix}$$

$$(sI - A)^{-1} = \frac{1}{s^2 + 4s + 4}\begin{bmatrix} s+4 & 1 \\ -4 & s \end{bmatrix}$$

$$H(s) = \begin{bmatrix} -2 & -3 \end{bmatrix}\frac{1}{s^2 + 4s + 4}\begin{bmatrix} s+4 & 1 \\ -4 & s \end{bmatrix}\begin{bmatrix} 0 \\ 1 \end{bmatrix} + 1 = \frac{s^2 + s + 2}{s^2 + 4s + 4}$$

Figure 10.9 shows the generation of the output $y(t)$ from input $x(t)$ for the repeated poles only. Diagonal canonical form is applicable only for real roots. If complex-conjugate poles appear, they should be separately implemented in some other canonical form.

Cascade and Parallel Realizations

As the order of the system becomes high, the canonical realizations tend to more sensitive to small parameter variations due to the tendency of the poles and zeros to occur in clusters. Consequently, in practice, a higher-order system is usually decomposed into first- and second-order sections connected in cascade or parallel. In the cascade form, the transfer function is decomposed into a product of first- and second-order transfer functions.

$$H(s) = H_1(s)H_2(s) \cdots H_m(s)$$

The performance of the system varies with the groupings of the numerators and denominators and has to be optimized with some respect. In cascade realization, one simple pole should be implemented first with the given input, and subsequently, the next simple pole should be implemented again with the output of the first section as the input, and so on. The complex-conjugate poles should be realized by a second-order section, for implementation feasibility in practice.

In the parallel form, the transfer function is decomposed into a sum of first- and second-order transfer functions.

$$H(s) = g + H_1(s) + H_2(s) +, \cdots, + H_m(s)$$

where g is a constant. In both cascade and parallel realizations, each section is independent and clustering of poles and zeros is avoided as the maximum number of poles and zeros in each section is limited to two.

10.5   Linear Transformation of State Vectors and Diagonalization

The linear transformation is also called similarity transformation. For a specific input–output relationship of a continuous system, the system can have different internal structures. By a linear transformation of a state vector, we can obtain another vector implying different internal structure of the system. Let the state-space model of a system be

$$\dot{q}(t) = Aq(t) + Bx(t)$$
$$y(t) = Cq(t) + Dx(t)$$

Let

$$q(t) = P\overline{q}(t)$$

where P is an invertible similarity transformation matrix.

$$\dot{\overline{q}}(t) = P^{-1}\dot{q}(t)$$
$$= P^{-1}(Aq(t) + Bx(t))$$
$$= P^{-1}(AP\overline{q}(t) + Bx(t))$$
$$= P^{-1}AP\overline{q}(t) + P^{-1}Bx(t)$$

With $\overline{A} = P^{-1}AP$ and $\overline{B} = P^{-1}B$, the state equation can be written as

$$\dot{\overline{q}}(t) = \overline{A}\overline{q}(t) + \overline{B}x(t)$$

With $\overline{C} = CP$, the output equation can be written as

$$y(t) = \overline{C}\overline{q}(t) + Dx(t)$$

Some properties of A and $\overline{A}$ matrices can be used to check the computation of $\overline{A}$. The determinants of A and $\overline{A}$ are equal. The determinants of $(s\boldsymbol{I} - A)$ and $(s\boldsymbol{I} - \overline{A})$ are the same. The traces (the sum of the diagonal elements) of A and $\overline{A}$ are equal.

The transformation matrix $\boldsymbol{P}$ to get the diagonal form from the controllable canonical form is

$$
P = \begin{bmatrix}
1 & 1 & \cdots & 1 \\
\lambda_1 & \lambda_2 & \cdots & \lambda_N \\
\lambda_1^2 & \lambda_2^2 & \cdots & \lambda_N^2 \\
\vdots & \vdots & & \vdots \\
\lambda_1^{N-1} & \lambda_2^{N-1} & \cdots & \lambda_N^{N-1}
\end{bmatrix}
$$

where λ_i, $i = 1, 2, \ldots N$ for an $N \times N$ matrix with all distinct roots. For a root λ_r repeated m times, the first column corresponding to that root will remain the same as for a distinct root and the $m - 1$ successive columns will be successive derivatives of the first column with respect to λ_r.

Example 10.11 Transform the state-space model of the given system using the transformation matrix

$$
P = \begin{bmatrix} 1 & 0 \\ -2 & 1 \end{bmatrix}
$$

$$
A = \begin{bmatrix} 0 & 1 \\ -4 & -4 \end{bmatrix}, \quad B = \begin{bmatrix} 0 \\ 1 \end{bmatrix}, \quad C = \begin{bmatrix} -2 & -3 \end{bmatrix}, \quad D = 0
$$

Verify that the transfer function remains the same with either state-space model. Verify that the determinants and traces of A and $\overline{A}$ are the same.

Solution

$$
P_* = \begin{bmatrix} 1 & 0 \\ -2 & 1 \end{bmatrix}, \quad P^{-1} = \begin{bmatrix} 1 & 0 \\ 2 & 1 \end{bmatrix}
$$

$$
\overline{A} = P^{-1}AP = \begin{bmatrix} 1 & 0 \\ 2 & 1 \end{bmatrix}\begin{bmatrix} 0 & 1 \\ -4 & -4 \end{bmatrix}\begin{bmatrix} 1 & 0 \\ -2 & 1 \end{bmatrix} = \begin{bmatrix} -2 & 1 \\ 0 & -2 \end{bmatrix}
$$

$$
\overline{B} = P^{-1}B = \begin{bmatrix} 1 & 0 \\ 2 & 1 \end{bmatrix}\begin{bmatrix} 0 \\ 1 \end{bmatrix} = \begin{bmatrix} 0 \\ 1 \end{bmatrix}
$$

$$
\overline{C} = CP = \begin{bmatrix} -2 & -3 \end{bmatrix}\begin{bmatrix} 1 & 0 \\ -2 & 1 \end{bmatrix} = \begin{bmatrix} 4 & -3 \end{bmatrix}
$$

The transfer function of the system, using either state-space model, is the same.

$$
H(s) = \frac{-3s - 2}{s^2 + 4s + 4}
$$

∎

Example 10.12 Transform the state-space model of the given system using the transformation matrix.

$$A = \begin{bmatrix} 0 & 1 \\ -2 & -3 \end{bmatrix}, \quad B = \begin{bmatrix} 0 \\ 1 \end{bmatrix}, \quad C = \begin{bmatrix} 3 & 1 \end{bmatrix}, \quad D = 0$$

Find the two eigenvalues, λ_1 and λ_2, of matrix A. Then, the transformation matrix is given by

$$P = \begin{bmatrix} 1 & 1 \\ \lambda_1 & \lambda_2 \end{bmatrix}$$

Verify that the transfer function remains the same with either state-space model. Verify that the determinants and traces of A and $\overline{A}$ are the same.

Solution

$$P = \begin{bmatrix} 1 & 1 \\ -1 & -2 \end{bmatrix}, \quad P^{-1} = \begin{bmatrix} 2 & 1 \\ -1 & -1 \end{bmatrix}$$

$$\overline{A} = P^{-1}AP = \begin{bmatrix} 2 & 1 \\ -1 & -1 \end{bmatrix}\begin{bmatrix} 0 & 1 \\ -2 & -3 \end{bmatrix}\begin{bmatrix} 1 & 1 \\ -1 & -2 \end{bmatrix} = \begin{bmatrix} -1 & 0 \\ 0 & -2 \end{bmatrix}$$

$$\overline{B} = P^{-1}B = \begin{bmatrix} 2 & 1 \\ -1 & -1 \end{bmatrix}\begin{bmatrix} 0 \\ 1 \end{bmatrix} = \begin{bmatrix} 1 \\ -1 \end{bmatrix}$$

$$\overline{C} = CP = \begin{bmatrix} 3 & 1 \end{bmatrix}\begin{bmatrix} 1 & 1 \\ -1 & -2 \end{bmatrix} = \begin{bmatrix} 2 & 1 \end{bmatrix}$$

The transfer function of the system, using either state-space model, is the same.

$$H(s) = \begin{bmatrix} 2 & 1 \end{bmatrix} \frac{1}{s^2 + 3s + 2}\begin{bmatrix} s+2 & 0 \\ 0 & s+1 \end{bmatrix}\begin{bmatrix} 1 \\ -1 \end{bmatrix}$$

$$= \frac{s+3}{s^2 + 3s + 2}$$

∎

Eigenvectors

Let A be an $N \times N$ matrix. For some nonzero column vector

$$v = \begin{bmatrix} v_1 \\ v_2 \\ \vdots \\ v_N \end{bmatrix}$$

if

$$Av = \lambda v \qquad \text{or} \qquad (A - \lambda I)v = 0$$

for some scalar λ, called the eigenvalue of A, then $v \neq 0$ is called the eigenvector associated with λ. I is the $N \times N$ identity matrix.

Example 10.13 Diagonalize the matrix

$$A = \begin{bmatrix} 5 & 1 \\ 4 & 2 \end{bmatrix}$$

Verify that the determinants and traces of A and its diagonalized version are the same.

Solution A square matrix with all its elements zero, except on the main diagonal, is a diagonal matrix. The characteristic equation is

$$\begin{vmatrix} 5 - \lambda & 1 \\ 4 & 2 - \lambda \end{vmatrix} = \lambda^2 - 7\lambda + 6 = 0$$

Solving, we get $\lambda_1 = 1$ and $\lambda_2 = 6$. Let us find the two eigenvectors using their definition. For $\lambda_1 = 1$, we get

$$(5 - 1)x + y = 0 = 4x + y$$
$$4x + (2 - 1)y = 0 = 4x + y$$

Assuming $x = 1$, we get $y = -4$. The first eigenvector is

$$\begin{bmatrix} 1 \\ -4 \end{bmatrix}$$

Similarly, the second eigenvector is

$$\begin{bmatrix} 1 \\ 1 \end{bmatrix}$$

The transformation matrix and its inverse are

$$P = \begin{bmatrix} 1 & 1 \\ -4 & 1 \end{bmatrix} \quad \text{and} \quad P^{-1} = \begin{bmatrix} 0.2 & -0.2 \\ 0.8 & 0.2 \end{bmatrix}$$

$$D = P^{-1}AP = \begin{bmatrix} 1 & 0 \\ 0 & 6 \end{bmatrix}$$

Computation of the State-Transition Matrix e^{At} Using Diagonalization

In this method, A matrix is diagonalized first. The procedure is illustrated in computing the state-transition matrix for the A matrix in Problem 10.5.

$$A = \begin{bmatrix} -1 & -2 \\ \frac{10}{3} & 0 \end{bmatrix}$$

$$\lambda_1 = -0.5 + j2.5331 \qquad \text{and} \qquad \lambda_2 = -0.5 - j2.5331$$

For $\lambda_1 = (-0.5 + j2.5331)$, we get

$$(-1 - (-0.5 + j2.5331))x - 2y = 0 \text{ or } y = -0.2500 - j1.2666x$$
$$\tfrac{10}{3}x + (0.5 - j2.5331)y = 0$$

Solving for x and y, and since the second eigenvector is the conjugate of the first one, we get the transformation matrix and its inverse as

$$P = \begin{bmatrix} 1 & 1 \\ -0.2500 - j1.2666 & -0.2500 + 1.2666 \end{bmatrix} \quad \text{and}$$

$$P^{-1} = \begin{bmatrix} 0.5 + j0.0987 & j0.3948 \\ 0.5 - j0.0987 & -j0.3948 \end{bmatrix}$$

$$D = P^{-1}AP = \begin{bmatrix} -0.5 + j2.5331 & 0 \\ 0 & -0.5 - j2.533 \end{bmatrix}$$

$$e^{At} = P \begin{bmatrix} e^{(-0.5+j2.5331)t} & 0 \\ 0 & e^{(-0.5-j2.533)t} \end{bmatrix} P^{-1}$$

$$= \begin{bmatrix} 1.0193e^{-0.5t}\cos(2.5331t + 0.1949) & -0.7896e^{-0.5t}\cos(2.5331t - 0.5\pi) \\ 1.3160e^{-0.5t}\cos(2.5331t - 0.5\pi) & 1.0193e^{-0.5t}\cos(2.5331t - 0.1949) \end{bmatrix}$$

as obtained earlier.

Example 10.14 Let

$$A = \begin{bmatrix} 0 & 1 \\ -4 & -4 \end{bmatrix}$$

Find P such that

$$J = P^{-1}AP$$

is a Jordan matrix. Verify that the determinants and traces of A and its Jordan form are the same.

Solution The characteristic equation is

$$\begin{vmatrix} 0 - \lambda & 1 \\ -4 & -4 - \lambda \end{vmatrix} = \lambda^2 + 4\lambda + 4 = 0$$

Solving, we get $\lambda_1 = -2$ and $\lambda_2 = -2$. The Jordan matrix is

$$J = \begin{bmatrix} -2 & 1 \\ 0 & -2 \end{bmatrix}$$

For $\lambda_1 = 2$, we get

$$(0 + 2)x + y = 0 = 2x + y$$
$$-4x + (-4 + 2)y = 0 = 2x + y$$

Arbitrarily assuming $x = 1$, we get $y = -2$. The first eigenvector is

$$\begin{bmatrix} 1 \\ -2 \end{bmatrix}$$

For the second eigenvector, from the relationship $AP = PJ$, we get

$$2x + y = 1$$

Arbitrarily assuming $x = 0$, we get $y = 1$. The second eigenvector is

$$\begin{bmatrix} 0 \\ 1 \end{bmatrix}$$

Ensure that the two eigenvectors are linearly independent, so that P is invertible. The transformation matrix and its inverse are

$$P = \begin{bmatrix} 1 & 0 \\ -2 & 1 \end{bmatrix} \quad \text{and} \quad P^{-1} = \begin{bmatrix} 1 & 0 \\ 2 & 1 \end{bmatrix}$$

$$J = P^{-1}AP = \begin{bmatrix} -2 & 1 \\ 0 & -2 \end{bmatrix}$$

10.6 Controllability

A system is completely state controllable if each initial state $q(t_0)$ at any t_0 can be transferred to any final state $q(t_f)$ in a finite time, $t_f > t_0$, by means of an unconstrained control input vector $x(t)$. It implies that all the states are affected by the input. The presence of common factors on the input and output sides of the transform of the differential equation is an indication of uncontrollability and/or unobservability. While the internal and external descriptions are identical for most systems, there are some systems whose external description is inadequate to represent the system exactly. This happens when the system is uncontrollable and/or unobservable. One of the necessary and sufficient conditions for complete controllability of a system is that the rank of the controllability matrix

$$[B \mid AB \mid A^2 B \mid \cdots \mid A^{N-1} B]$$

be N, where A and B are the system and input matrices, respectively. Of course, if the rank of the controllability matrix is N, then its determinant must be nonzero.

Another test for controllability is that $P^{-1} B$ has all nonzero elements, after the system has been transformed to diagonal form. It is assumed that all the eigenvalues of A matrix are distinct. For systems with repeated poles, the entries of $P^{-1} B$ in any last row of each Jordan block and the entries corresponding to the all the distinct poles are not all zero is the condition for controllability in diagonalized form.

Example 10.15 Determine the controllability of the system, described by

$$A = \begin{bmatrix} 0 & -1 \\ 2 & -3 \end{bmatrix}, \quad B = \begin{bmatrix} 1 \\ 0 \end{bmatrix}$$

Solution

$$(sI - A) = s \begin{bmatrix} 1 & 0 \\ 0 & 1 \end{bmatrix} - \begin{bmatrix} 0 & -1 \\ 2 & -3 \end{bmatrix} = \begin{bmatrix} s & 1 \\ -2 & s+3 \end{bmatrix}$$

The characteristic polynomial of the system, given by the determinant of this matrix, is

$$s^2 + 3s + 2 = (s + 1)(s + 2)$$

The roots of this equation are the two eigenvalues $\{\lambda_1 = -1, \lambda_2 = -2\}$. For finding the eigenvectors, we use the equation

$$(\lambda I - A) v = 0$$

For $\lambda = -1$, we get

$$\begin{bmatrix} -1 - 0 & 1 \\ -2 & -1 + 3 \end{bmatrix} \begin{bmatrix} v(0) \\ v(1) \end{bmatrix} = 0$$

We get

$$-v(0) + v(1) = 0$$

$$-2v(0) + 2v(1) = 0$$

As these equations are dependent, we take one of them and give any nontrivial solution. For example, $\{v(0) = 1, v(1) = 1\}$. For $\lambda = -2$, we get

$$\begin{bmatrix} -2-0 & 1 \\ -2 & -2+3 \end{bmatrix} \begin{bmatrix} v(0) \\ v(1) \end{bmatrix} = 0$$

Similarly, $\{v(0) = 1, v(1) = 2\}$. Therefore,

$$P = \begin{bmatrix} 1 & 1 \\ 1 & 2 \end{bmatrix}$$

The first and second columns are, respectively, the eigenvectors corresponding to eigenvalues -1 and -2.

$$P^{-1}AP = \begin{bmatrix} 2 & -1 \\ -1 & 1 \end{bmatrix} \begin{bmatrix} 0 & -1 \\ 2 & -3 \end{bmatrix} \begin{bmatrix} 1 & 1 \\ 1 & 2 \end{bmatrix} = \Lambda = \begin{bmatrix} -1 & 0 \\ 0 & -2 \end{bmatrix}$$

$$P^{-1}B = \begin{bmatrix} 2 \\ -1 \end{bmatrix}$$

has no zero entries, and therefore, the system is controllable.

The rank of the matrix

$$[B \mid AB] = \begin{bmatrix} 1 & 0 \\ 0 & 2 \end{bmatrix}$$

is 2. Therefore, the system is controllable.   ∎

With

$$B = \begin{bmatrix} 1 \\ 1 \end{bmatrix}$$

the rank of the matrix

$$[B \mid AB] = \begin{bmatrix} 1 & -1 \\ 1 & -1 \end{bmatrix}$$

is 1. Therefore, the system is uncontrollable.

$$P^{-1}B = \begin{bmatrix} 1 \\ 0 \end{bmatrix}$$

has one zero entry, and therefore, the system is uncontrollable.

10.7 Observability

A system is completely observable if every initial state $q(t_0)$ can be exactly determined from the output values over a finite interval of time $t_0 \leq t \leq t_f$. It implies that all the states affect the output. One of the necessary and sufficient conditions for complete observability of a system is that the rank of the observability matrix

$$[C^T \mid A^T C^T \mid (A^T)^2 C^T \mid \cdots \mid (A^T)^{N-1} C^T]$$

is N, where A^T and C^T are the transposes of the system and output matrices, respectively. Of course, if the rank of the observability matrix is N, then its determinant must be nonzero. Another test for observability is that CP has all nonzero elements, after the system has been transformed to diagonal form.

Example 10.16 Determine the observability of the system, described by

$$A = \begin{bmatrix} 3 & 2 \\ 1 & 4 \end{bmatrix}, \quad C = \begin{bmatrix} 1 & 1 \end{bmatrix}$$

Solution

$$(sI - A) = s \begin{bmatrix} 1 & 0 \\ 0 & 1 \end{bmatrix} - \begin{bmatrix} 3 & 2 \\ 1 & 4 \end{bmatrix} = \begin{bmatrix} s - 3 & -2 \\ -1 & s - 4 \end{bmatrix}$$

The characteristic polynomial of the system, given by the determinant of this matrix, is

$$s^2 - 7s + 10 = (s - 2)(s - 5)$$

The roots of this equation are the two eigenvalues $\{\lambda_1 = 2, \lambda_2 = 5\}$. For finding the eigenvectors, we use the equation

$$(\lambda I - A)v = 0$$

For $\lambda = 2$, we get

$$\begin{bmatrix} 2 - 3 & -2 \\ -1 & 2 - 4 \end{bmatrix} \begin{bmatrix} v(0) \\ v(1) \end{bmatrix} = 0$$

We get

$$-v(0) - 2v(1) = 0$$
$$-v(0) - 2v(1) = 0$$

As these equations are dependent, we take one of them and give any nontrivial solution. For example, $\{v(0) = 2, v(1) = -1\}$. For $\lambda = 5$, we get

$$\begin{bmatrix} 5-3 & -2 \\ -1 & 5-4 \end{bmatrix} \begin{bmatrix} v(0) \\ v(1) \end{bmatrix} = 0$$

Similarly, for $\lambda = 2$, $\{v(0) = 1, v(1) = 1\}$. Therefore,

$$P = \begin{bmatrix} 2 & 1 \\ -1 & 1 \end{bmatrix}$$

The first and second columns are, respectively, the eigenvectors corresponding to eigenvalues 2 and 5.

$$P^{-1}AP = \begin{bmatrix} \frac{1}{3} & -\frac{1}{3} \\ \frac{1}{3} & \frac{2}{3} \end{bmatrix} \begin{bmatrix} 3 & 2 \\ 1 & 4 \end{bmatrix} \begin{bmatrix} 2 & 1 \\ -1 & 1 \end{bmatrix} = \Lambda = \begin{bmatrix} 2 & 0 \\ 0 & 5 \end{bmatrix}$$

$$CP = \begin{bmatrix} 1 & 2 \end{bmatrix}$$

has no zero entries, and therefore, the system is observable.

The rank of the matrix

$$[C^T \mid A^T C^T] = \begin{bmatrix} 1 & 4 \\ 1 & 6 \end{bmatrix}$$

is 2. Therefore, the system is observable.   ∎

With

$$C = \begin{bmatrix} 1 & -1 \end{bmatrix}$$

the rank of the matrix

$$[C^T \mid A^T C^T] = \begin{bmatrix} 1 & 2 \\ -1 & -2 \end{bmatrix}$$

is 1. Therefore, the system is unobservable.

$$CP = \begin{bmatrix} 3 & 0 \end{bmatrix}$$

has a zero entry, and therefore, the system is unobservable.

10.8 Summary

- The state-space model of a system gives the complete description of the dynamical behavior of a system. This is achieved by including the internal states, called the state variables, of the system in addition to the input and output variables.
- The general state-space model description, with $x(t)$ input and $y(t)$ output and assuming single input and single output system, is given by

$$\dot{q}(t) = Aq(t) + Bx(t)$$

$$y(t) = Cq(t) + Dx(t)$$

where A is the state matrix, B is the input matrix, C is the output matrix, and D is the transmission matrix. The first equation is always a set of first-order differential equations, called the state equation. This equation involves the state, its first derivative, and the input. The second equation is the output equation. The output is expressed as a linear combination of the state variables and the input.
- The Laplace transform of the zero-input and zero-state components of the state vector $q(t)$ is

$$Q(s) = \overbrace{(sI - A)^{-1}q(0^-)}^{q_{zi}(s)} + \overbrace{(sI - A)^{-1}BX(s)}^{q_{zs}(s)}$$

The inverse Laplace transforms of the first and the second expressions on the right-hand side are, respectively, the zero-input and zero-state components of the state vector $q(t)$. Taking the Laplace transform of the output equation, we get

$$Y(s) = CQ(s) + DX(s)$$

- With the system initial conditions zero, the transfer function is given by

$$H(s) = \frac{Y(s)}{X(s)} = (C(sI - A)^{-1}B + D)$$

- The characteristic polynomial of a system is given by the determinant of the matrix $(sI - A)$, where I is the identity matrix of the same size as A.

-

$$q(t) = \overbrace{e^{At}q(0^-)}^{q_{zi}(t)} + \overbrace{\int_{0^-}^{t} e^{A(t-\tau)}Bx(\tau)d\tau}^{q_{zs}(t)}$$

The first and the second expressions on the right-hand side are, respectively, the zero-input and zero-state components of the state vector $q(t)$.

- The matrix e^{At} is called the state-transition matrix or the fundamental matrix of the system. The inverse Laplace transform of $((sI - A)^{-1})$ is e^{At}.
- While the transfer function of a system is unique, it can be realized in several ways.
- For a specific input–output relationship of a continuous system, the system can have different internal structures. By a linear transformation of a state matrix, we can obtain another matrix implying different internal structure of the system.
- A square matrix with all its elements zero, except on the main diagonal, is a diagonal matrix.
- A system is controllable if there always exists a control input $x(t)$ that transfers any state of the system to any other state in a finite interval of time.
- A system is said to be observable if it is possible to determine the state of a system from its inputs and outputs in a finite interval of time.

Exercises

10.1 The input voltage is the unit-step function, $u(t)$. Assume that the series circuit consists of the resistor with value $R = 3\Omega$ and the inductor with value $L = 2H$, shown in Fig. 10.10. Let the initial current through the inductor be zero. Find the current through the circuit after the excitation is applied: (i) by transfer function analysis, (ii) by state-space analysis in the time domain, and (iii) by state-space analysis in the frequency domain. Verify that all the 3 methods yield the same results.

10.2 The input voltage is the unit-step function, $u(t)$. Assume that the series circuit consists of the resistor with value $R = 3\Omega$, capacitor with value $C = 2F$, and the inductor with value $L = 1H$, shown in Fig. 10.11. Let the initial conditions be zero. Find the current through the circuit and the voltage across the capacitor after the excitation is applied: (i) by transfer function analysis, (ii) by state-space analysis in the time domain, and (iii) by state-space analysis in the frequency domain. Verify that all the 3 methods yield the same results.

* 10.3 Find the voltage v_1 across the inductor in the circuit shown, in both time and frequency domains in Fig. 10.12, using the state-space analysis. Let the initial conditions be zero.

Fig. 10.10 An RL circuit in the time domain

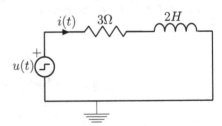

Fig. 10.11 An *RLC* circuit in time domain

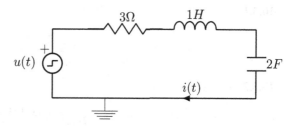

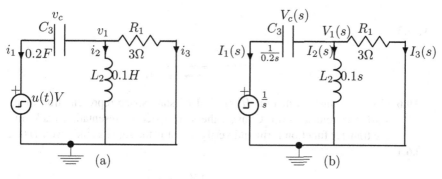

Fig. 10.12 A circuit with unit-step input voltage, in the time domain on the left and in the frequency domain on the right

10.4 Given a transfer function $H(s)$, find its state-space representation of its controllable canonical form. Convert the state-space representation back into the transfer function form and verify that it is the same as the given $H(s)$.

* 10.4.1

$$H(s) = \frac{1}{s^2 + 3s + 2}$$

10.4.2

$$H(s) = \frac{s + 1}{s^2 + s + 2}$$

10.4.3

$$H(s) = \frac{2s^2 + s + 1}{s^2 + 2s + 1}$$

10.5 Given a transfer function $H(s)$, find its state-space representation of its diagonal canonical form. Convert the state-space representation back into the transfer function form and verify that it is the same as the given $H(s)$.

10.5.1

$$H(s) = \frac{2s^2 + s + 1}{s^2 + 3s + 2}$$

* 10.5.2

$$H(s) = \frac{s + 3}{s^2 + 2s}$$

10.5.3

$$H(s) = \frac{1}{s^2 + 5s + 6}$$

10.6 Given a transfer function $H(s)$, find its state-space representation of its Jordan canonical form. Convert the state-space representation back into the transfer function form and verify that it is the same as the given $H(s)$.

10.6.1

$$H(s) = \frac{2s^2 + s + 1}{s^2 + 2s + 1}$$

10.6.2

$$H(s) = \frac{s + 1}{s^2 + 4s + 4}$$

* 10.6.3

$$H(s) = \frac{2s + 3}{s^2 + 6s + 9}$$

10.7 Determine the controllability of the system, described by

$$A = \begin{bmatrix} 3 & 2 \\ 1 & 4 \end{bmatrix}, \quad B = \begin{bmatrix} 0 \\ 1 \end{bmatrix}$$

Use both the transformation method and the controllability matrix method and verify that the conclusion is the same.

* 10.8 Determine the controllability of the system, described by

$$A = \begin{bmatrix} 3 & 2 \\ 1 & 4 \end{bmatrix}, \quad B = \begin{bmatrix} 1 \\ 1 \end{bmatrix}$$

Use both the transformation method and the controllability matrix method and verify that the conclusion is the same.

* **10.9** Determine the observability of the system, described by

$$A = \begin{bmatrix} 1 & 2 \\ 1 & 4 \end{bmatrix}, \quad C = \begin{bmatrix} 1 & 1 \end{bmatrix}$$

Use both the transformation method and the observability matrix method and verify that the conclusion is the same.

10.10 Determine the observability of the system, described by

$$A = \begin{bmatrix} 2 & 0 \\ 0 & 4 \end{bmatrix}, \quad C = \begin{bmatrix} 0 & 1 \end{bmatrix}$$

Use both the transformation method and the observability matrix method and verify that the conclusion is the same.

Chapter 11
Design of Control Systems in State Space

In earlier chapters, we designed control systems using root locus and Bode plot. Using the state-space representation, more versatile control systems can be designed. The necessary and sufficient condition for arbitrary pole-placement is that the system is completely state controllable. The state variables are available for feedback. Control input is unconstrained. The closed-loop poles should lie at desired locations. In root locus method of design, only dominant poles are located. In pole-placement design, all the poles are located as desired.

Figure 11.1 shows the block diagram representation of the state-space model of a closed-loop system with state feedback and with single output. The system has no input. The purpose of the system is to recover from any disturbances and maintain its zero-input condition. This type of systems with zero or nonzero constant input is called regulator systems.

Consider the state and output equations of a control system

$$\dot{q} = Aq + Bx$$
$$y = Cq + Dx$$

Let the control signal be

$$x = -Kq$$

Then,

$$\dot{q} = (A - BK)q$$
$$y = (C - DK)q$$

Now, the characteristic equation of the feedback system is

$$\det(sI - (A - BK)) = 0$$

Comparing this equation with that of the desired poles, we get the values of K. The desired poles are chosen from experience and expected to give acceptable

D. Sundararajan, *Control Systems*, https://doi.org/10.1007/978-3-030-98445-8_11

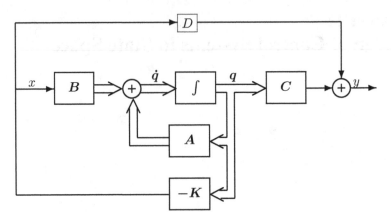

Fig. 11.1 Block diagram representation of the state-space model of a closed-loop system with state feedback

transient response. The dimensions of the variables are: (1) state vector q with N state variables; (2) scalar control and output signals; (3) $N \times N$ state matrix A; (4) $N \times 1$ input matrix B; (5) $1 \times N$ output matrix C; (6) D scalar; (7) gain matrix K $1 \times N$ vector.

11.1   Design by Pole-Placement

The desired poles are chosen first, based on the experience using root locus method. A pair of dominant poles is chosen. Then, other poles are located far from these poles. The next step is to determine the appropriate feedback gain matrix. We present three methods for the determination of the state feedback gain matrix K.

11.1.1   Direct Comparison Method

Example 11.1 The state and input matrices of a regulator are

$$A = \begin{bmatrix} 0 & 1 \\ -2 & -3 \end{bmatrix}, \quad B = \begin{bmatrix} 0 \\ 1 \end{bmatrix}$$

The desired pole locations are at $-2 \pm j2$. Design the system with state feedback by determining the gain matrix K. Let the initial state vector due to some disturbance be

$$q(0^-) = \begin{bmatrix} 1 \\ 0.5 \end{bmatrix}$$

Verify that the responses to the initial condition approach zero as $t \to \infty$.

Solution The system is state controllable. The desired characteristic equation is $s^2 + 4s + 8 = 0$. Let the gain matrix be

$$K = [k_1 \quad k_2]$$

Then,

$$
\begin{aligned}
|sI - A + BK| &= \left| \begin{bmatrix} s & 0 \\ 0 & s \end{bmatrix} - \begin{bmatrix} 0 & 1 \\ -2 & -3 \end{bmatrix} + \begin{bmatrix} 0 \\ 1 \end{bmatrix} [k_1 \ k_2] \right| \\
&= \begin{vmatrix} s & -1 \\ 2 + k_1 & s + 3 + k_2 \end{vmatrix} \\
&= s^2 + s(3 + k_2) + (2 + k_1) = 0
\end{aligned}
$$

Comparing with $s^2 + 4s + 8 = 0$, we get

$$K = [k_1 \quad k_2] = [6 \quad 1]$$

This method is simple but can be used only for lower-order systems in practice. Let $C = [1 \ 0]$. Figure 11.2 shows the block diagram of the system.

Let us find the zero-input response to the initial condition.

$$
\begin{aligned}
|sI - A + BK| &= \left| \begin{bmatrix} s & 0 \\ 0 & s \end{bmatrix} - \begin{bmatrix} 0 & 1 \\ -2 & -3 \end{bmatrix} + \begin{bmatrix} 0 \\ 1 \end{bmatrix} [6 \ 1] \right| \\
&= \begin{vmatrix} s & -1 \\ 8 & s + 4 \end{vmatrix}
\end{aligned}
$$

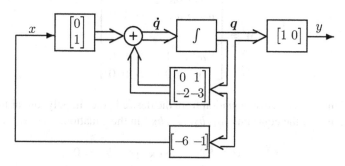

Fig. 11.2 Block diagram representation of the state-space model of a closed-loop system with state feedback and single output

$$(sI - A)^{-1} = \frac{1}{s^2 + 4s + 8} \begin{bmatrix} s + 4 & 1 \\ -8 & s \end{bmatrix} = \begin{bmatrix} \frac{s+4}{s^2+4s+8} & \frac{1}{s^2+4s+8} \\ \frac{-8}{s^2+4s+8} & \frac{s}{s^2+4s+8} \end{bmatrix}$$

$$q_{zi}(s) = (sI - A)^{-1}q(0^-) = \begin{bmatrix} \frac{s+4}{s^2+4s+8} & \frac{1}{s^2+4s+8} \\ \frac{-8}{s^2+4s+8} & \frac{s}{s^2+4s+8} \end{bmatrix} \begin{bmatrix} 1 \\ 0.5 \end{bmatrix} = \begin{bmatrix} \frac{s+4.5}{s^2+4s+8} \\ \frac{0.5s-8}{s^2+4s+8} \end{bmatrix}$$

Taking the inverse Laplace transform, we get

$$q_{zi}(t) = \begin{bmatrix} (1.6008e^{-2t}\cos(2t - 0.8961))u(t) \\ (4.5277e^{-2t}\cos(2t + 1.4601))u(t) \end{bmatrix}$$

11.1.2   Using Transformation Matrix

First, check whether the system is completely state controllable. If so, determine the characteristic equation of the system and find the values of the constants $\{a_1, a_2, \ldots, a_N\}$ in the equation

$$|sI - A| = s^N + a_1 s^{N-1} + \cdots + a_{N-1}s + a_N$$

Next, find the transformation matrix T that transforms the system equations into controllable canonical form. If the system equations are already in controllable canonical form, $T = I$. Otherwise, $T = MW$. M is the controllability matrix

$$M = [B \mid AB \mid \cdots \mid A^{N-1}B]$$

and

$$W = \begin{bmatrix} a_{N-1} & a_{N-2} & \cdots & a_1 & 1 \\ a_{N-2} & a_{N-3} & \cdots & 1 & 0 \\ \vdots & \vdots & & \vdots & \vdots \\ a_1 & 1 & \cdots & 0 & 0 \\ 1 & 0 & \cdots & 0 & 0 \end{bmatrix}$$

Next, find the characteristic equation from the desired poles in polynomial form and find the values of the constants $\{b_1, b_2, \ldots, b_N\}$ in the equation

$$s^N + b_1 s^{N-1} + \cdots + b_{N-1}s + b_N = 0$$

Now, the gain matrix is given by

$$K = \begin{bmatrix} (b_N - a_N) & (b_{N-1} - a_{N-1}) & \cdots & (b_2 - a_2) & (b_1 - a_1) \end{bmatrix} T^{-1}$$

Example 11.2 The state and input matrices of a regulator are

$$A = \begin{bmatrix} 0 & -2 \\ 1 & -3 \end{bmatrix}, \quad B = \begin{bmatrix} -2 \\ -3 \end{bmatrix}$$

The desired pole locations are at $-1 \pm j0.85$. Design the system with state feedback by determining the gain matrix K. Let the initial state vector due to some disturbance be

$$q(0^-) = \begin{bmatrix} 1 \\ -1 \end{bmatrix}$$

Verify that the responses to the initial condition approach zero as $t \to \infty$.

Solution First, we have to find controllability matrix.

$$M = [B \mid AB \mid \cdots \mid A^{N-1}B]$$

For this example,

$$M = [B \mid AB] = \begin{bmatrix} -2 & 6 \\ -3 & 7 \end{bmatrix}$$

The determinant of M is nonzero and arbitrary pole-placement is possible. The characteristic equation of the given system is $s^2 + 3s + 2 = s^2 + a_1 s + a_2 = 0$.

$$W = \begin{bmatrix} 3 & 1 \\ 1 & 0 \end{bmatrix}$$

The transformation matrix and its inverse are

$$T = MW = \begin{bmatrix} 0 & -2 \\ -2 & -3 \end{bmatrix}, \quad T^{-1} = \begin{bmatrix} 0.75 & -0.5 \\ -0.5 & 0 \end{bmatrix}$$

The desired characteristic equation is $s^2 + 2s + 1.7225 = s^2 + b_1 s + b_2 = 0$. The gain matrix is

$$K = [(b_2 - a_2) \ (b_1 - a_1)]T^{-1} = [-0.2775 \ -1] \begin{bmatrix} 0.75 & -0.5 \\ -0.5 & 0 \end{bmatrix} = [0.2919 \ 0.1387]$$

Let us find the zero-input response to the initial condition.

$$|sI - A + BK| = \left| \begin{bmatrix} s & 0 \\ 0 & s \end{bmatrix} - \begin{bmatrix} 0 & -2 \\ 1 & -3 \end{bmatrix} + \begin{bmatrix} -2 \\ -3 \end{bmatrix} [0.2919 \ 0.1387] \right|$$

$$= \begin{vmatrix} s - 0.5838 & 1.7225 \\ -1.8756 & s + 2.5838 \end{vmatrix}$$

$$(s\boldsymbol{I} - \boldsymbol{A})^{-1} = \frac{1}{s^2 + 2s + 1.7225} \begin{bmatrix} s + 2.5838 & -1.7225 \\ 1.8756 & s - 0.5838 \end{bmatrix}$$

$$= \begin{bmatrix} \frac{s+2.5838}{s^2+2s+1.7225} & \frac{-1.7225}{s^2+2s+1.7225} \\ \frac{1.8756}{s^2+2s+1.7225} & \frac{s-0.5838}{s^2+2s+1.7225} \end{bmatrix}$$

$$\boldsymbol{q}_{zi}(s) = (s\boldsymbol{I} - \boldsymbol{A})^{-1}\boldsymbol{q}(0^-) = \begin{bmatrix} \frac{s+2.5838}{s^2+2s+1.7225} & \frac{-1.7225}{s^2+2s+1.7225} \\ \frac{1.8756}{s^2+2s+1.7225} & \frac{s-0.5838}{s^2+2s+1.7225} \end{bmatrix} \begin{bmatrix} 1 \\ -1 \end{bmatrix}$$

$$= \begin{bmatrix} \frac{s+4.3063}{s^2+2s+1.7225} \\ \frac{-s+2.4594}{s^2+2s+1.7225} \end{bmatrix}$$

Taking the inverse Laplace transform, we get

$$\boldsymbol{q}_{zi}(t) = \begin{bmatrix} (4.0163e^{-t}\cos(0.85t - 1.3192))u(t) \\ (4.1909e^{-t}\cos(0.85t - 1.8117))u(t) \end{bmatrix}$$

11.1.3 Using Ackermann's Formula

Assume that the system is completely state controllable. The characteristic equation from the desired poles is

$$s^N + b_1 s^{N-1} + \cdots + b_{N-1}s + b_N = 0$$

Then, from Cayley–Hamilton theorem,

$$\phi(\boldsymbol{A}) = \boldsymbol{A}^N + b_1\boldsymbol{A}^{N-1} + \cdots + b_{N-1}\boldsymbol{A} + b_N\boldsymbol{I}$$

For a Nth order system, the gain matrix is given by

$$\boldsymbol{K} = [0\ 0\ \cdots\ 1][\boldsymbol{B} \mid \boldsymbol{AB} \mid \cdots \mid \boldsymbol{A}^{N-1}\boldsymbol{B}]^{-1}\phi(\boldsymbol{A})$$

Example 11.3 The state and input matrices of a regulator are

$$\boldsymbol{A} = \begin{bmatrix} 0 & 1 \\ -2 & -3 \end{bmatrix}, \quad \boldsymbol{B} = \begin{bmatrix} 0 \\ 1 \end{bmatrix}$$

The desired pole locations are at $-2 \pm j2$. Design the system with state feedback by determining the gain matrix K.

Solution The controllability matrix and its inverse are

$$\begin{bmatrix} 0 & 1 \\ 1 & -3 \end{bmatrix} \quad \begin{bmatrix} 3 & 1 \\ 1 & 0 \end{bmatrix}$$

The desired characteristic equation is $s^2 + 4s + 8 = 0$.

$$\phi(A) = A^2 + 4A + 8I$$

$$= \begin{bmatrix} -2 & -3 \\ 6 & 7 \end{bmatrix} + 4\begin{bmatrix} 0 & 1 \\ -2 & -3 \end{bmatrix} + 8\begin{bmatrix} 1 & 0 \\ 0 & 1 \end{bmatrix} = \begin{bmatrix} 6 & 1 \\ -2 & 3 \end{bmatrix}$$

$$K = \begin{bmatrix} 0 & 1 \end{bmatrix}\begin{bmatrix} 3 & 1 \\ 1 & 0 \end{bmatrix}\begin{bmatrix} 6 & 1 \\ -2 & 3 \end{bmatrix} = \begin{bmatrix} 1 & 0 \end{bmatrix}\begin{bmatrix} 6 & 1 \\ -2 & 3 \end{bmatrix} = \begin{bmatrix} 6 & 1 \end{bmatrix}$$

Transformation and Ackermann method are preferred for higher-order systems, since all matrix computations can be carried out by a computer.

11.2 State Observers

In pole-placement design of control systems, access to the state variables is necessary. However, if they are not accessible then we are forced to estimate them. A device that estimates the state variables from the control and output variables is called the state observer. The state-space representation of a system is given by

$$\dot{q}(t) = Aq + Bx$$

$$y(t) = Cq$$

The necessary and sufficient condition for observability of the states is that the system is completely observable. The same three methods used to find the controller gain matrix are applicable to find the state observer gain matrix K_e with the B matrix replaced by the C matrix.

Direct Comparison Method

Example 11.4 The state and output matrices of a regulator are

$$A = \begin{bmatrix} 0 & -2 \\ 1 & -3 \end{bmatrix}, \quad C = \begin{bmatrix} 0 & 1 \end{bmatrix}$$

The desired pole locations are at $p_1 = -12$ and $p_2 = -8$. Design the full-order state observer by determining the gain matrix K_c.

Solution The system is already in observable canonical form. The observability matrix is

$$[C^T \mid A^T C^T] = \begin{bmatrix} 0 & 1 \\ 1 & -3 \end{bmatrix}$$

The determinant of this matrix is nonzero. The desired characteristic equation is $s^2 + 20s + 96 = 0$. Let the gain matrix be

$$K_c = \begin{bmatrix} k_{c1} \\ k_{c2} \end{bmatrix}$$

Then, the characteristic equation of the observer becomes

$$|sI - A + K_c C| = \left| \begin{bmatrix} s & 0 \\ 0 & s \end{bmatrix} - \begin{bmatrix} 0 & -2 \\ 1 & -3 \end{bmatrix} + \begin{bmatrix} k_{c1} \\ k_{c2} \end{bmatrix} [0 \ 1] \right|$$

$$= \left| \begin{bmatrix} s & 2 + k_{c1} \\ -1 & s + 3 + k_{c2} \end{bmatrix} \right|$$

$$= s^2 + s(3 + k_{c2}) + (2 + k_{c1}) = 0$$

Comparing with $s^2 + 20s + 96 = 0$, we get

$$K_c = \begin{bmatrix} k_{c1} \\ k_{c2} \end{bmatrix} = \begin{bmatrix} 94 \\ 17 \end{bmatrix}$$

This method is simple but can be used only for lower-order systems in practice.

Using the Transformation Matrix T
First, check whether the system is completely state observable. If so, determine the characteristic equation of the system and find the values of the constants $\{a_1, a_2, \ldots, a_N\}$ in the equation

$$|sI - A| = s^N + a_1 s^{N-1} + \cdots + a_{N-1}s + a_N = 0$$

Next, find the transformation matrix Q that transforms the system into observable canonical form. If the system equations are already in observable canonical form, $Q = I$. Otherwise, $Q = (WN^T)^{-1}$. N is the observability matrix

$$N = [C^T \mid A^T C^T \mid \cdots \mid (A)^{N-1} C^T]$$

W matrix is the same as defined for regulator design. Next, find the desired characteristic equation from the desired poles in polynomial form and find the values of the constants $\{b_1, b_2, \ldots, b_N\}$ in the equation

$$s^N + b_1 s^{N-1} + \cdots + b_{N-1} s + b_N = 0$$

Now, the gain matrix is given by

$$K_c = Q \begin{bmatrix} (b_N - a_N) \\ (b_{N-1} - a_{N-1}) \\ \vdots \\ (b_2 - a_2) \\ (b_1 - a_1) \end{bmatrix}$$

Example 11.5 The state and output matrices of a regulator are

$$A = \begin{bmatrix} 0 & -2 \\ 1 & -3 \end{bmatrix}, \quad C = \begin{bmatrix} 1 & 3 \end{bmatrix}$$

The desired poles are $p_1 = -9$ and $p_2 = -6$. Design the state observer by determining the gain matrix K_c.

Solution The system is not in observable canonical form. The observability matrix is

$$N = [C^T \mid A^T C^T] = \begin{bmatrix} 1 & 3 \\ 3 & -11 \end{bmatrix}$$

The determinant of this matrix is nonzero.

$$W = \begin{bmatrix} 3 & 1 \\ 1 & 0 \end{bmatrix}$$

The transformation matrix and its inverse are

$$T = W N^T = \begin{bmatrix} 6 & -2 \\ 1 & 3 \end{bmatrix}, \quad T^{-1} = Q = \begin{bmatrix} 00.15 & 0.10 \\ -0.05 & 0.30 \end{bmatrix}$$

The characteristic equation of the given system is $s^2 + 3s + 2 = s^2 + a_1 s + a_2 = 0$. The desired characteristic equation is $s^2 + 15s + 54 = s^2 + b_1 s + b_2 = 0$. The gain matrix is

$$K = Q \begin{bmatrix} (b_2 - a_2) \\ (b_1 - a_1) \end{bmatrix} = \begin{bmatrix} 0.15 & 0.10 \\ -0.05 & 0.30 \end{bmatrix} \begin{bmatrix} 54 - 2 \\ 15 - 3 \end{bmatrix} = \begin{bmatrix} 9 \\ 1 \end{bmatrix}$$

Using Ackermann's Formula

The characteristic equation from the desired poles is

$$s^N + b_1 s^{N-1} + \cdots + b_{N-1} s + b_N = 0$$

Then,

$$\phi(A) = A^N + b_1 A^{N-1} + \cdots + b_{N-1} A + b_N I$$

For a Nth order system, the gain matrix is given by

$$K_c = \phi(A)[C^T \mid A^T C^T \mid \cdots \mid (A)^{N-1} C^T]^{-1} \begin{bmatrix} 0 \\ 0 \\ \vdots \\ 1 \end{bmatrix}$$

Example 11.6 The state and input matrices of a regulator are

$$A = \begin{bmatrix} 0 & -2 \\ 1 & -3 \end{bmatrix}, \quad C = [0 \; 1]$$

The desired pole locations are at $p_1 = -12$ and $p_1 = -8$. Design the state observer by determining the gain matrix K_c.

Solution The observability matrix and its inverse are

$$N = [C^T \mid A^T C^T] = \begin{bmatrix} 0 & 1 \\ 1 & -3 \end{bmatrix} \quad \text{and} \quad N^{-1} \begin{bmatrix} 3 & 1 \\ 1 & 0 \end{bmatrix}$$

The desired characteristic equation is $s^2 + 20s + 96 = 0$.

$$\phi(A) = A^2 + 20A + 96I$$

$$= \begin{bmatrix} -2 & 6 \\ -3 & 7 \end{bmatrix} + 20 \begin{bmatrix} 0 & -2 \\ 1 & -3 \end{bmatrix} + 96 \begin{bmatrix} 1 & 0 \\ 0 & 1 \end{bmatrix} = \begin{bmatrix} 94 & -34 \\ 17 & 43 \end{bmatrix}$$

$$K_c = \begin{bmatrix} 94 & -34 \\ 17 & 43 \end{bmatrix} \begin{bmatrix} 3 & 1 \\ 1 & 0 \end{bmatrix} \begin{bmatrix} 0 \\ 1 \end{bmatrix} = \begin{bmatrix} 94 \\ 17 \end{bmatrix}$$

Transformation and Ackermann method are preferred for higher-order systems, since all matrix computations can be carried out by a computer.

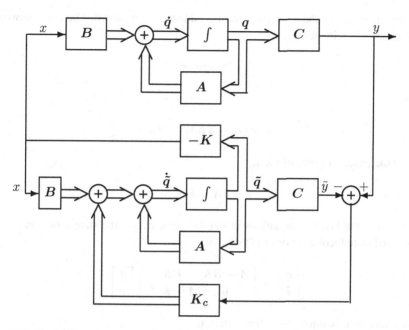

Fig. 11.3 Block diagram representation of the state-space model of a closed-loop system with observer-based state feedback

11.2.1   Design of Regulator Systems with Observers

When the states are not available for feedback, it becomes a necessity to employ an observer to estimate the states. Figure 11.3 shows the block diagram representation of the state-space model of a closed-loop system with observer-based state feedback and with single output. Consider a system with complete controllability and observability characterized by the state-space representation

$$\dot{q} = Aq + Bx$$
$$y = Cq + Dx$$

with $D = 0$. We found the gain matrix for the system and observer independently in earlier examples. Now, we have to compute both the gain matrices and find the transfer function of the combined system. With the observed state variables $\tilde{q}$, the state feedback is

$$x = -K\tilde{q}$$

Now, the state equation is approximated to

$$\dot{q} = Aq - BK\tilde{q} = (A - BK)q + BK(q - \tilde{q})$$

On the right side of the equation a term is added and subtracted. The error between the exact and estimated states is

$$e = (q - \tilde{q})$$

Then,

$$\dot{q} = (A - BK)q + BKe$$

From the observer point of view,

$$\dot{e} = (A - K_cC)e$$

Expressing the last two equations in matrix form, we get the state equation of the observed state feedback control system as

$$\begin{bmatrix} \dot{q} \\ \dot{e} \end{bmatrix} = \begin{bmatrix} A - BK & BK \\ 0 & A - K_cC \end{bmatrix} \begin{bmatrix} q \\ e \end{bmatrix}$$

The characteristic equation of the system is

$$\begin{vmatrix} sI - A + BK & -BK \\ 0 & sI - A + K_cC \end{vmatrix} = 0$$

Evaluating the determinant, we get

$$|sI - A + BK||sI - A + K_cC| = 0$$

The poles are the desired ones and those of the observer only.

Transfer Function of the Observer-Based Controller

Consider the state-space description of a completely observable system

$$\dot{q} = Aq + Bx$$
$$y = Cq + Dx$$

with $D = 0$. With the observed state variables $\tilde{q}$, the state feedback is

$$x = -K\tilde{q}$$

The state-space description of a relaxed observer system is

$$\dot{\tilde{q}} = (A - K_cC - BK)\tilde{q} + K_cy$$

$$x = -K\tilde{q}$$

There are three inputs that determine $\dot{\tilde{q}}$. Taking the Laplace transform of the state equation and solving for $\tilde{Q}(s)$, we get

$$\tilde{Q}(s) = (sI - A + K_cC + BK)^{-1}K_cY(s)$$

Substituting the expression $\tilde{Q}(s)$ in the Laplace transform of $x = -K\tilde{q}$, we get

$$Y_{co}(s) = -K(sI - A + K_cC + BK)^{-1}K_cY(s)$$

Therefore, the transfer function of the observer-controller is

$$\frac{Y_{co}(s)}{Y(s)} = -K(sI - A + K_cC + BK)^{-1}K_c$$

This transfer function is the controller for the system. The negative sign is taken care in the negative feedback configuration, as show in Fig. 11.4.

Example 11.7 The state, input, and output matrices of a system are

$$A = \begin{bmatrix} 0 & 1 \\ -2 & -3 \end{bmatrix}, \quad B = \begin{bmatrix} 0 \\ 1 \end{bmatrix}, \quad C = \begin{bmatrix} 1 & 0 \end{bmatrix}$$

Let the desired poles of the system are at $p_1 = -2 + j2$ and $p_2 = -2 - j2$ and those of the observer are at $p_{o1} = -12$ and $p_{o2} = -8$. Design an observer-based regulator the system.

Solution The controllability matrix of the system is

$$\begin{bmatrix} 0 & 1 \\ 1 & -3 \end{bmatrix}$$

The observability matrix of the system is

$$\begin{bmatrix} 1 & 0 \\ 0 & 1 \end{bmatrix}$$

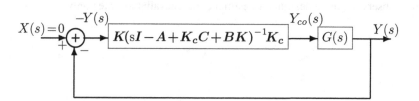

$X(s) = 0$ $-Y(s)$ $K(sI - A + K_cC + BK)^{-1}K_c$ $Y_{co}(s)$ $G(s)$ $Y(s)$

Fig. 11.4 Block diagram of a system with observer-controller

The determinant of both the matrices are nonzero. The state feedback gain matrix for the system is

$$K = [k_1 \quad k_2] = [6 \quad 1]$$

The state feedback gain matrix for the observer is

$$K_c = \begin{bmatrix} k_{c1} \\ k_{c2} \end{bmatrix} = \begin{bmatrix} 17 \\ 43 \end{bmatrix}$$

The transfer function of the observer-controller is

$$G_c(s) = \begin{bmatrix} 6 & 1 \end{bmatrix} \begin{bmatrix} s+17 & -1 \\ 51 & s+4 \end{bmatrix}^{-1} \begin{bmatrix} 17 \\ 43 \end{bmatrix} = \frac{145s + 530}{s^2 + 21s + 119}$$

The transfer function of the system is

$$G(s) = \frac{1}{s^2 + 3s + 2}$$

The closed-loop transfer function of the observer-based system is

$$H(s) = \frac{145s + 530}{s^4 + 24s^3 + 184s^2 + 544s + 768}$$

The roots are

$$\{-2 + j2, \ -2 - j2\}, \quad \{-12, \ -8\}$$

which are the desired poles and observer poles, as expected.

Let the initial conditions be

$$x(0) = \begin{bmatrix} 1 \\ 1 \end{bmatrix}, \qquad e(0) = \begin{bmatrix} 0.5 \\ -1 \end{bmatrix}$$

Figure 11.5a and b show the response of the state variables to initial conditions. Figure 11.5c and d show the response of the error state variables to initial conditions. If some state values of the system are available for measurement, then a minimum-order observer can be designed to estimate the unavailable states only.

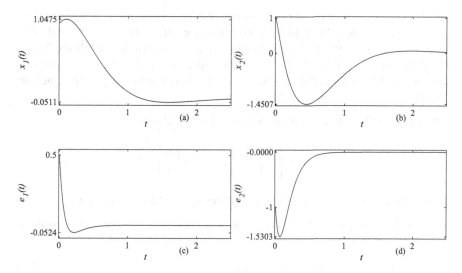

Fig. 11.5 Response to initial conditions

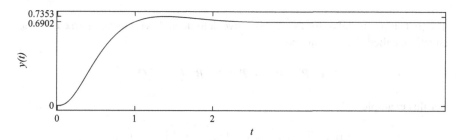

Fig. 11.6 Unit-step response

11.3   Design of Control Systems with Observers

We just add the input to find the response of the observer-based system. The unit-step response for the system designed in the last section is shown in Fig. 11.6. Different response can be obtained by changing the desired pole locations of the system and those of the observer. The controller can also be placed in the feedback path.

11.3.1   *Quadratic Optimal Regulator Systems*

We know that overshoot and settling time conflict each other. The use of controllers enables us to control these parameters, as shown in the design of control systems in earlier chapters. This type of problems can also be formulated as a mathematical optimization problem by defining a cost function that must be minimized. The objective in control systems is to reduce the errors in the response. While the theory of optimization is quite complex, using this approach to determine the optimum gain vector for state variable feedback is easier with a software package, such as Matlab©.

Example 11.8 The state and input matrices of a regulator are

$$A = \begin{bmatrix} 0 & 1 \\ -2 & -3 \end{bmatrix}, \quad B = \begin{bmatrix} 0 \\ 1 \end{bmatrix}$$

Design the system with state feedback by determining the gain matrix K.

Solution Four matrices, A, B, Q, R, are involved in the optimization process. Typically,

$$Q = \begin{bmatrix} 1 & 0 \\ 0 & 1 \end{bmatrix}, \quad R = 1$$

The first step is to solve for the real positive definite and symmetric matrix P in the equation, called Riccati equation,

$$A^T P + P A - P B R^{-1} B^T P = -Q$$

For this example,

$$\begin{bmatrix} 0 & -2 \\ 1 & -3 \end{bmatrix} \begin{bmatrix} p_{11} & p_{12} \\ p_{12} & p_{22} \end{bmatrix} + \begin{bmatrix} p_{11} & p_{12} \\ p_{12} & p_{22} \end{bmatrix} \begin{bmatrix} 0 & 1 \\ -2 & -3 \end{bmatrix}$$

$$- \begin{bmatrix} p_{11} & p_{12} \\ p_{12} & p_{22} \end{bmatrix} \begin{bmatrix} 0 \\ 1 \end{bmatrix} [1] \begin{bmatrix} 0 & 1 \end{bmatrix} \begin{bmatrix} p_{11} & p_{12} \\ p_{12} & p_{22} \end{bmatrix} = - \begin{bmatrix} 1 & 0 \\ 0 & 1 \end{bmatrix}$$

The matrix equality represents a set of three simultaneous equations. Using the possible solutions to these equations, we have to reconstruct a positive definite matrix. For this example, the matrix, using Matlab©, is

$$P = \begin{bmatrix} 1.2361 & 0.2361 \\ 0.2361 & 0.2361 \end{bmatrix}$$

One test for a symmetric 2×2 matrix

$$\begin{bmatrix} a & b \\ b & d \end{bmatrix}$$

is positive definite is that

$$a > 0 \text{ and } ad - b^2 > 0$$

The optimal feedback gain matrix is given as

$$\boldsymbol{K} = R^{-1}\boldsymbol{B}^T\boldsymbol{P} = [1][0\ \ 1]\begin{bmatrix} 1.2361 & 0.2361 \\ 0.2361 & 0.2361 \end{bmatrix} = [0.2361\ \ 0.2361]$$

The poles of the closed-loop transfer function are located at $-1.6180 \pm j0.7862$.

11.4 Digital Implementation of Continuous-Time Systems

Due to advances in digital system technology and fast numerical algorithms, digital implementation of systems is in widespread use. However, most practical systems are of continuous nature. Therefore, these systems are characterized by the transfer function in the s-domain. The time variable varies continuously and the frequency varies from $-\infty$ to ∞. For the implementation of practical systems, however, samples of the time-domain signal and finite bandwidth are adequate. That is, the response of practical systems can be considered as both time-limited and bandlimited with adequate accuracy. No physical system can generate a frequency of infinite order. The conclusion is that all practical systems can be implemented with sampled-data with adequate accuracy. The digital implementation provides the advantages of low cost and high reliability. The sampling theorem states that the sampling frequency must be greater than twice the highest frequency of interest. In practice, due to the nonideal behavior of physical systems, a slightly higher sampling frequency is necessary than that specified by the sampling theorem.

11.4.1 The Bilinear Transformation

We[1] have designed, in earlier chapters, the transfer functions of systems in the s-domain. Now, these transfer functions have to be transformed to the z-domain for digital implementation of systems. That is, we have to derive a suitable transformation for replacing the variable s of the continuous-time transfer function by the digital system variable z. While there are some methods for the transformation, the transformation more often used in practice is called the bilinear (linear with respect to each of two variables) transformation. It is based on transforming a differential

---

[1] Adapted from my book Digital Signal Processing—An Introduction, Springer, 2021 with permission.

equation into a difference equation. This transformation is also called Tustin transformation. Using this transformation, a continuous-time transfer function $H(s)$ is converted to its corresponding discrete-time transfer function $H(z)$, so that the system can be implemented using digital systems. The transformation from the s-domain to the z-domain is given by $z = e^{sT_s}$, where T_s is the sampling interval. The Tustin transformation is a linear approximation to this relation.

Consider the first-order transfer function in the s-domain

$$H(s) = \frac{Y(s)}{X(s)} = \frac{1}{s+1}$$

The corresponding differential equation is

$$\frac{dy(t)}{dt} + y(t) = x(t)$$

While the derivative term can be approximated by a finite difference, in the bilinear transformation, the differential equation is integrated and the integral is approximated by the trapezoidal formula for numerical integration. In numerical integration, the area to be integrated is divided into subintervals and an approximation function is used to find the area enclosed in each subinterval. One of the approximation functions often used is based on the trapezoid. Let the sampling interval be T_s. Then, the trapezoidal rule is

$$y(n) = y(n-1) + \frac{T_s}{2}(x(n) + x(n-1))$$

The z-transform of this equation, in the transfer function form, is

$$H(z) = \frac{Y(z)}{X(z)} = \frac{T_s}{2}\frac{(z+1)}{(z-1)}$$

Since $1/s$ is integration in the s-domain, we have to make the substitution

$$s = \frac{2}{T_s}\frac{(z-1)}{(z+1)} \tag{11.1}$$

in the transfer function $H(s)$ of the system to get an equivalent discrete transfer function $H(z)$. Applying this transformation, called the bilinear transformation, to the transfer function of the first-order system, we get

$$H(z) = \frac{Y(z)}{X(z)} = \frac{T_s(z+1)}{(T_s+2)z + (T_s-2)} \tag{11.2}$$

Frequency Warping

The frequency range in the s-plane is infinite. That is, $-\infty \leq \omega_a \leq \infty$. The effective frequency range in the z-plane is finite and it is periodic. That is, $-\pi < \omega_d T_s \leq \pi$. We have to find the relationship between ω_a and ω_d in the digital implementation.

Let $s = (\sigma + j\omega_a)$ and $z = re^{j\omega_d T_s}$. Substituting for s and z in the bilinear transformation formula, we get

$$s = (\sigma + j\omega_a) = \frac{2}{T_s}\frac{(z-1)}{(z+1)} = \frac{2}{T_s}\frac{(re^{j\omega_d T_s} - 1)}{(re^{j\omega_d T_s} + 1)}$$

Equating the real and imaginary parts of both sides, we get

$$\sigma = \frac{2}{T_s}\left(\frac{r^2 - 1}{1 + r^2 + 2r\cos(\omega_d T_s)}\right)$$

and

$$\omega_a = \frac{2}{T_s}\left(\frac{2r\sin(\omega_d T_s)}{1 + r^2 + 2r\cos(\omega_d T_s)}\right)$$

The real part $\sigma < 0$ if $r < 1$ and $\sigma > 0$ if $r > 1$. That is, the left-half of the s plane maps into the inside of the unit-circle and the right-half maps into the outside. If $r = 1$, $\sigma = 0$, the $j\omega$ axis maps on the unit-circle. With $r = 1$, we get

$$\omega_a = \frac{2}{T_s}\left(\frac{\sin(\omega_d T_s)}{1 + \cos(\omega_d T_s)}\right) = \frac{2}{T_s}\tan\left(\frac{T_s}{2}\omega_d\right)$$

using trigonometric half-angle formula. The relationship between the frequency variables ω_d and ω_a is given as

$$\omega_a = \frac{2}{T_s}\tan\left(\frac{T_s}{2}\omega_d\right) \quad \text{and} \quad \omega_d = \frac{2}{T_s}\tan^{-1}\left(\frac{T_s}{2}\omega_a\right) \tag{11.3}$$

Figure 11.7 shows the relationship between ω_a and ω_d with $T_s = 1$ second. As an infinite range is mapped to a finite range, the relation between ω_d and ω_a is highly nonlinear, except for a short range.

The frequency responses of the continuous and discrete ($T_s = 0.1$) transfer functions of the first-order system are shown in Fig. 11.8. The shapes of the responses of the two transfer functions are similar and they are identical for some range. At some point, the discrete response shows higher attenuation due to the warping effect. For example, the attenuations of the two transfer functions at $\omega = 15$ radians/second are, respectively, -23.5411 dB and -25.4176 dB in Fig. 11.8a. If the

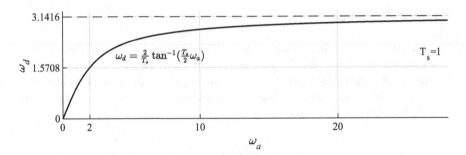

Fig. 11.7 Mapping between ω_a and ω_d in the bilinear transformation

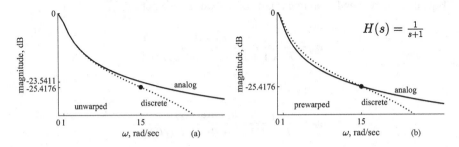

Fig. 11.8 The frequency response; (**a**) without prewarping; (**b**) with prewarping

discrete transfer function should have the same attenuation at a particular frequency $\omega_d = 15$ radians, then the corresponding analog frequency must be, with $T_s = 0.1$,

$$\omega_a = (2/0.1)\tan((0.1/2)15) = 18.6319$$

The frequency response after prewarping is shown in Fig. 11.8b.

Application of the Bilinear Transformation

Filters of any order are usually, for analytical or implementation advantages, decomposed into a set of first- and second-order sections. The formulas relating the analog and digital second-order transfer functions can be derived. Let the first- and second-order analog transfer functions be

$$H(s) = \frac{c_1 s + c_0}{d_1 s + d_0} \quad \text{and} \quad H(s) = \frac{c_2 s^2 + c_1 s + c_0}{d_2 s^2 + d_1 s + d_0}$$

The transformation is

$$s = \frac{2}{T_s}\frac{(z-1)}{(z+1)} = k\frac{(z-1)}{(z+1)}$$

Then, the corresponding digital second-order transfer function is

$$H(z) = \frac{\frac{(c_0+c_1k+c_2k^2)}{D}z^2 + \frac{2(c_0-c_2k^2)}{D}z + \frac{(c_0-c_1k+c_2k^2)}{D}}{z^2 + \frac{2(d_0-d_2k^2)}{D}z + \frac{(d_0-d_1k+d_2k^2)}{D}}$$

where $D = (d_0 + d_1k + d_2k^2)$.

For the first-order analog transfer function, the corresponding digital transfer function is

$$H(z) = \frac{\frac{(c_0+c_1k)}{D}z + \frac{(c_0-c_1k)}{D}}{z + \frac{(d_0-d_1k)}{D}}$$

where $D = (d_0 + d_1k)$

Example 11.9 The transfer function of a system is

$$H(s) = \frac{s^2 + s + 2}{s^2 + 2s + 1}$$

The sampling interval is $T_s = 0.1$ sec. Find the corresponding discrete-time transfer function $H(z)$.

Repeat the problem with $T_s = 0.3$ s.

The unit-step response computed from $H(s)$ is

$$y(t) = (2 - e^{-t} - 2te^{-t})u(t)$$

Using the transformation formulas, we get, for $T_s = 0.1$,

$$H(z) = \frac{0.9569z^2 - 1.805z + 0.8662}{z^2 - 1.81z + 0.8186}$$

For $T_s = 0.3$,

$$H(z) = \frac{0.9036z^2 - 1.444z + 0.6767}{z^2 - 1.478z + 0.5463}$$

Figure 11.9a and b show the response for unit-step input signal with $T_s = 0.1$ sec and $T_s = 0.3$ sec, respectively, along with that of the continuous-time system. In figure (a), the responses are almost identical, while the responses are slightly different due the increase of the sampling interval in figure (b). In discrete-time systems, the sampling interval is a critical parameter and has to be selected appropriately.

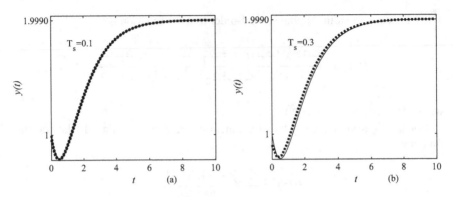

Fig. 11.9 Unit-step response of the system: (**a**) $T_s = 0.1$ sec; (**b**) $T_s = 0.3$ sec

11.5 Summary

- Using the state-space representation, more versatile control systems can be designed.
- The necessary and sufficient condition for arbitrary pole-placement is that the system is completely state controllable. The state variables are available for feedback.
- In pole-placement design, all the poles are located as desired.
- Regulator system has no input. The purpose of the system is to recover from any disturbances and maintain its zero or constant input condition.
- The desired poles are chosen from experience and expected to give acceptable transient response.
- Three methods are commonly used for the determination of the state feedback gain matrix K. One of the three methods is the Ackermann's formula.
- A device that estimates the state variables from the control and output variables is called the state observer.
- When the states are not available for feedback, it becomes a necessity to employ an observer to estimate the states.
- We just add the input to find the response of the observer-based system to find the response.
- Regulator design can also be formulated as a mathematical optimization problem by defining a cost function that must be minimized.
- Due to advances in digital system technology and fast numerical algorithms, digital implementation of systems is in widespread use.
- All practical systems can be implemented with sampled-data with adequate accuracy.
- The transfer functions of systems in the s-domain have to be transformed to the z-domain for digital implementation of systems. That is, we have to derive a suitable transformation for replacing the variable s of the continuous-time transfer function by the digital system variable z.

- Using Tustin or bilinear transformation, a continuous-time transfer function $H(s)$ is converted to its corresponding discrete-time transfer function $H(z)$. The frequency range in the s-plane is infinite. That is, $-\infty \le \omega_a \le \infty$. The effective frequency range in the z-plane is finite and it is periodic. That is, $-\pi < \omega_d T_s \le \pi$. We have to find the relationship between ω_a and ω_d in the digital implementation.
- As an infinite range is mapped to a finite range, the relation between ω_d and ω_a is highly nonlinear, except for a short range. This creates the warping effect.

Exercises

11.1 The state and input matrices of a regulator are

$$A = \begin{bmatrix} 0 & 1 \\ -2 & -1 \end{bmatrix}, \quad B = \begin{bmatrix} 0 \\ 1 \end{bmatrix}$$

The desired pole locations are at $-1 \pm j2$. Design the system with state feedback by determining the gain matrix K using the direct comparison method. Let the initial state vector due to some disturbance be

$$q(0^-) = \begin{bmatrix} 1 \\ 1 \end{bmatrix}$$

Verify that the responses to the initial condition approach zero as $t \to \infty$.

* 11.2 The state and input matrices of a regulator are

$$A = \begin{bmatrix} 0 & 1 \\ -3 & -4 \end{bmatrix}, \quad B = \begin{bmatrix} 0 \\ 1 \end{bmatrix}$$

The desired pole locations are at $-3 \pm j3$. Design the system with state feedback by determining the gain matrix K using the direct comparison method. Let the initial state vector due to some disturbance be

$$q(0^-) = \begin{bmatrix} 2 \\ 1 \end{bmatrix}$$

Verify that the responses to the initial condition approach zero as $t \to \infty$.

11.3 The state and input matrices of a regulator are

$$A = \begin{bmatrix} 0 & 1 \\ -2 & -2 \end{bmatrix}, \quad B = \begin{bmatrix} 0 \\ 1 \end{bmatrix}$$

The desired pole locations are at $-3 \pm j3$. Design the system with state feedback by determining the gain matrix K using the direct comparison method. Let the initial state vector due to some disturbance be

$$q(0^-) = \begin{bmatrix} 1 \\ 1 \end{bmatrix}$$

Verify that the responses to the initial condition approach zero as $t \to \infty$.

* **11.4** The state and input matrices of a regulator are

$$A = \begin{bmatrix} 0 & -1 \\ 1 & -1 \end{bmatrix}, \quad B = \begin{bmatrix} 1 \\ 1 \end{bmatrix}$$

The desired pole locations are at $-1 \pm j1$. Design the system with state feedback by determining the gain matrix K using the transformation matrix. Let the initial state vector due to some disturbance be

$$q(0^-) = \begin{bmatrix} 1 \\ -1 \end{bmatrix}$$

Verify that the responses to the initial condition approach zero as $t \to \infty$.

11.5 The state and input matrices of a regulator are

$$A = \begin{bmatrix} 0 & -2 \\ 1 & -3 \end{bmatrix}, \quad B = \begin{bmatrix} 1 \\ 0 \end{bmatrix}$$

The desired pole locations are at $-2 \pm j2$. Design the system with state feedback by determining the gain matrix K using the transformation matrix. Let the initial state vector due to some disturbance be

$$q(0^-) = \begin{bmatrix} 1 \\ 1 \end{bmatrix}$$

Verify that the responses to the initial condition approach zero as $t \to \infty$.

* **11.6** The state and input matrices of a regulator are

$$A = \begin{bmatrix} 0 & 0 \\ 1 & -3 \end{bmatrix}, \quad B = \begin{bmatrix} 1 \\ 1 \end{bmatrix}$$

The desired pole locations are at $-1 \pm j1$. Design the system with state feedback by determining the gain matrix K using the transformation matrix. Let the initial state vector due to some disturbance be

$$q(0^-) = \begin{bmatrix} 0.5 \\ 1 \end{bmatrix}$$

Verify that the responses to the initial condition approach zero as $t \to \infty$.

11.7 The state and input matrices of a regulator are

$$A = \begin{bmatrix} 0 & 1 \\ -3 & -5 \end{bmatrix}, \quad B = \begin{bmatrix} 0 \\ 1 \end{bmatrix}$$

The desired pole locations are at $-2 \pm j1$. Design the system with state feedback by determining the gain matrix K using Ackermann's formula.

* 11.8 The state and input matrices of a regulator are

$$A = \begin{bmatrix} 0 & 1 \\ -1 & -1 \end{bmatrix}, \quad B = \begin{bmatrix} 1 \\ 1 \end{bmatrix}$$

The desired pole locations are at $-1 \pm j1$. Design the system with state feedback by determining the gain matrix K using Ackermann's formula.

11.9 The state and input matrices of a regulator are

$$A = \begin{bmatrix} 0 & 1 \\ -2 & -1 \end{bmatrix}, \quad B = \begin{bmatrix} 1 \\ 1 \end{bmatrix}$$

The desired pole locations are at $-1 \pm j1$. Design the system with state feedback by determining the gain matrix K using Ackermann's formula.

* 11.10 The state and output matrices of a regulator are

$$A = \begin{bmatrix} 0 & -1 \\ 1 & -3 \end{bmatrix}, \quad C = \begin{bmatrix} 0 & 1 \end{bmatrix}$$

The desired pole locations are at $p_1 = -10$ and $p_2 = -7$. Design the full-order state observer by determining the gain matrix K_c. Use the direct comparison method.

11.11 The state and output matrices of a regulator are

$$A = \begin{bmatrix} 0 & -1 \\ 1 & -1 \end{bmatrix}, \quad C = \begin{bmatrix} 1 & 0 \end{bmatrix}$$

The desired pole locations are at $p_1 = -7$ and $p_2 = -5$. Design the full-order state observer by determining the gain matrix K_c. Use the direct comparison method.

* 11.12 The state and output matrices of a regulator are

$$A = \begin{bmatrix} 0 & -1 \\ 1 & -1 \end{bmatrix}, \quad C = \begin{bmatrix} 1 & 1 \end{bmatrix}$$

The desired poles are $p_1 = -7$ and $p_2 = -5$. Design the state observer by determining the gain matrix K_c. Use the transformation matrix method.

11.13 The state and output matrices of a regulator are

$$A = \begin{bmatrix} 2 & -1 \\ 1 & -1 \end{bmatrix}, \quad C = \begin{bmatrix} 1 & 1 \end{bmatrix}$$

The desired poles are $p_1 = -7$ and $p_2 = -5$. Design the state observer by determining the gain matrix K_c. Use the transformation matrix method.

* 11.14 The state and input matrices of a regulator are

$$A = \begin{bmatrix} 1 & -2 \\ 1 & -3 \end{bmatrix}, \quad C = \begin{bmatrix} 0 & 1 \end{bmatrix}$$

The desired pole locations are at $p_1 = -12$ and $p_1 = -8$. Design the state observer by determining the gain matrix K_c using the Ackermann's formula.

11.15 The state and input matrices of a regulator are

$$A = \begin{bmatrix} 1 & -2 \\ 1 & -3 \end{bmatrix}, \quad C = \begin{bmatrix} 1 & -1 \end{bmatrix}$$

The desired pole locations are at $p_1 = -12$ and $p_1 = -8$. Design the state observer by determining the gain matrix K_c using the Ackermann's formula.

* 11.16 The continuous-time transfer function of a system is

$$H(s) = \frac{2s + 3}{s + 2}$$

The sampling interval is $T_s = 0.1$ sec. Find the corresponding discrete-time transfer function $H(z)$ using the bilinear transformation.

11.17 The contiuous-time transfer function of a system is

$$H(s) = \frac{s^2 + 2s + 2}{s^2 + 3s + 1}$$

The sampling interval is $T_s = 0.1$ sec. Find the corresponding discrete-time transfer function $H(z)$ using the bilinear transformation.

* 11.18 The contiuous-time transfer function of a system is

$$H(s) = \frac{1}{(s + 1)(s^2 + 2s + 1)}$$

The sampling interval is $T_s = 0.1$ s. Find the corresponding discrete-time transfer function $H(z)$ using the bilinear transformation.

Answers to Selected Exercises

Chapter 1

1.2.1 Sample values are

$$\{-1.0000, 1.7321, 1.0000, -1.7321\}$$

The rectangular form is

$$x(t) = -\cos\left(\frac{2\pi}{8}t\right) + 1.7321 \sin\left(\frac{2\pi}{8}t\right)$$

$$t_{max} = 2.68$$

1.3.3 $c(t) = 2.9064 \cos(\frac{2\pi}{8}t + 0.7084)$. The sample values of the sinusoid $c(t) = a(t) + b(t)$ are

$$\{2.2071, 0.2235, -1.8910, -2.8978, -2.2071, -0.2235, 1.8910, 2.8978\}$$

1.4.2

$$x(t) = \frac{2}{2}\left(e^{j(\frac{2\pi}{8}t - \frac{\pi}{3})} + e^{-j(\frac{2\pi}{8}t - \frac{\pi}{3})}\right)$$

The sample values of $x(t)$ are

$$\{1.0000, 1.9319, 1.7321, 0.5176, -1.0000, -1.9319, -1.7321, -0.5176\}$$

D. Sundararajan, *Control Systems*, https://doi.org/10.1007/978-3-030-98445-8

Chapter 2

2.1.3

$$x(t) = tu(t-2) = ((t-2)u(t-2) + 2u(t-2)) \leftrightarrow \left(\frac{e^{-2s}}{s^2} + \frac{2e^{-2s}}{s}\right)$$

2.2.2

$$e^{-2t}\sin(3t)u(t) \leftrightarrow \frac{3}{(s+2)^2+9}$$

2.3.3
The poles are located at $s = -j$ and $s = j$. $x(t) = (\sin(t))u(t)$. The poles are located at $s = -j3$ and $s = j3$. $x(at) = (\sin(3t))u(t)$.

2.4.1
Since $tu(t) \leftrightarrow \frac{1}{s^2}$, $t^2u(t) \leftrightarrow -\frac{d}{ds}\left(\frac{1}{s^2}\right) = \frac{2}{s^3}$.

2.5.2

$$x(0^+) = \lim_{s\to\infty} s\frac{1}{s+4} = 1$$

$$x(\infty) = \lim_{s\to 0} s\frac{1}{(s+4)} = 0$$

2.7

$$x(t) = \left(0.5 - 0.3333e^{-t} + 0.8333e^{-4t}\right)u(t)$$

2.9

$$x(t) = \left(-te^{-t} + 2e^{-t} - 2e^{-2t}\right)u(t)$$

2.12

$$x(t) = \left(-e^{-(t-1)} + 2e^{-2(t-1)}\right)u(t-1)$$

2.13

$$y(t) = \left(2e^{-t} + 3e^{-2t}\right)u(t)$$

Chapter 3

3.2

$$i(t) = \left(\frac{1}{3}e^{-\frac{t}{3}}\right)u(t)$$

$$i(t) = \left(\frac{2}{3}e^{-\frac{t}{3}}\right)u(t)$$

3.4

$$i(t) = (3e^{-t} - e^{-0.5t})u(t)$$
$$i(t) = (-2e^{-t} + 2e^{-0.5t})u(t)$$

3.6

$$v_1(t) = (0.3948e^{-0.5t}\cos(2.5331t - 0.5\pi))u(t)$$
$$i_1(t) = (0.5096e^{-0.5t}\cos(2.5331t + 0.1949))u(t)$$
$$i_3(t) = 0.5(0.3948e^{-0.5t}\cos(2.5331t - 0.5\pi) - 1)u(t)$$
$$i_2(t) = (0.5 + 0.5096e^{-0.5t}\cos(2.5331t + 2.9467))u(t)$$

Chapter 4

4.1

$$I(s) = \frac{1}{2s} \leftrightarrow i(t) = 0.5u(t)A$$

4.3

$$i_C(t) = 0.5e^{-0.5t}u(t)$$

4.5

$$i(t) = (-0.866e^{-0.7887t} + 0.866e^{-0.2113t})u(t)$$

4.7

$$i_C(t) = \frac{1}{3}e^{-\frac{1}{6}t}u(t)$$

4.9

$$i(t) = (-e^{-t} + e^{-0.5t})u(t)$$

Chapter 5

5.2

$$y1(t) = (1.0101 - 0.0093e^{-2.0196t} - 1.0008e^{-0.9804t})u(t)$$

$$y2(t) = (0.0253 + 0.0238e^{-2.0196t} - 0.0491e^{-0.9804t}u(t)$$

The total output is $y(t) = y1(t) + y2(t)$.

5.4

$$y(t) = (0.2 + 5e^{-t} + 6.7420e^{-1.3t} \cos(0.3317t + 2.4063))u(t)$$

5.5

$$y(t) = 10(0.0994e^{-3.5562t} - 0.1818e^{-2t}$$
$$+ 0.0835e^{-0.2219t} \cos(0.6104t - 0.1630))u(t)$$

Chapter 6

6.1.1

$$y(t) = \left(\frac{1}{3} - \frac{1}{3}e^{-3t}\right)u(t)$$

As $t \to \infty$, the response tends to $1/3$, making the error $2/3$.

6.2.2

$$y(t) = (t - 0.5 + 0.7071e^{-t} \cos(t + 0.7854))u(t)$$

As $t \to \infty$, the response tends to $t - 0.5$, making the error 0.5.

6.3.3

$$y(t) = (0.5t^2 - 8 + 0.0072e^{-3.6849t} - 0.0197e^{-2.1559t}$$
$$+ 8.2119e^{-0.0796t} \cos(0.3457t - 0.2208))u(t)$$

As $t \to \infty$, the response tends to $0.5t^2 - 8$, making the error 8.

6.4.1

$$y(t) = (1 + 1.1547e^{-2t}\cos(3.4641t + 2.618))u(t)$$

$$\omega_n = \sqrt{16} = 4, \zeta = 2/4 = 0.5, t_s = 4/(\omega_n\zeta) = 2, \omega_d = 3.4641$$

$$M_p = e^{-\frac{\pi\zeta}{\sqrt{1-\zeta^2}}} = 0.163, t_p = \pi/\omega_d = 0.9069$$

Chapter 7

7.1 The loop transfer function is

$$\frac{4.2107}{(s^2 + 4s + 3)}$$

7.3 The loop transfer function is

$$\frac{5.9557s + 0.59557}{s^3 + 4.01s^2 + 3.04s + 0.03}$$

7.5 The loop transfer function is

$$50.5068\frac{(s + 3.7501)}{s^3 + 11.0001s^2 + 18.0002s}$$

7.7 The loop transfer function is

$$168.3460\frac{(s + 2.6884)(s + 3.0840)(s + 0.54)}{(s + 1)(s + 2)(s + 4)(s + 6.6526)(s + 7.1356)(s + 0.1)}$$

Chapter 8

8.1

$$KG_c(s)G(s) = \frac{(13.01s + 3.816)}{(s^3 + 4.033s^2 + 3.13s + 0.0978)}$$

8.3

$$KG_c(s)G(s) = (0.65)(6.3408)\left(\frac{s + 1.1365}{s + 7.2065}\right)\left(\frac{8.6761}{s(s + 1)}\right)$$

8.5

$$KG_c(s)G(s) = \frac{3.8s^2 + 22.04s + 31.96}{s^3 + 4s^2 + 3s}$$

8.7

$$KG_c(s)G(s) = \frac{10.5s^2 + 6.93s + 0.588}{s^4 + 4s^3 + 3s^2}$$

Chapter 9

9.1.3

$$Gm = \infty, \qquad Pm = 90.8° \text{ at } 7.24 \text{ rad/sec}$$

The system is stable. The roots of the closed-loop transfer function are -6.4142, -3.5858.

9.1.5

$$Gm = 14 \text{ dB at } 1 \text{ rad/sec}, \qquad Pm = 2.95° \text{ at } 0.419 \text{ rad/sec}$$

The system is stable. The roots of the closed-loop transfer function are -3.1823, -1.7952, $-0.0113 \pm j0.4182$.

9.1.9

$$Gm = 6.02 \text{ dB at } 1 \text{ rad/sec}, \qquad Pm = 21.4° \text{ at } 0.682 \text{ rad/sec}$$

The system is stable. The roots of the closed-loop transfer function are -1.7549, $-0.1226 \pm j0.7449$.

9.2.2 The system is unstable. The roots of the closed-loop transfer function are -0.8, 0.5.

9.2.4 The system is unstable. The roots of the closed-loop transfer function are $0.8440 \pm j3.3372$, -1.6879.

9.3.2

$$Gm = -0.857 \text{ dB at } 1.17 \text{ rad/sec}, \qquad Pm = -4.29° \text{ at } 1.25 \text{ rad/sec}$$

The system is unstable.

9.4.2 The approximate loop transfer function is

$$\frac{-2s + 6.667}{s^3 + 4.333s^2 + 3.333s}$$

$$Gm = -0.519 \text{ dB} \text{ at } 1.2 \text{ rad/sec}, \qquad Pm = -2.43° \text{ at } 1.25 \text{ rad/sec}$$

The system is unstable. The roots are -4.3765, $0.0217 \pm j1.2341$.

Chapter 10

10.3

$$v_1(t) = (1.007e^{-0.8333t} \cos(7.0218t + 0.1181))u(t)$$

10.4.1

$$A = \begin{bmatrix} 0 & 1 \\ -2 & -3 \end{bmatrix}, \quad B = \begin{bmatrix} 0 \\ 1 \end{bmatrix}, \quad C = \begin{bmatrix} 1 & 0 \end{bmatrix}, \quad D = 0$$

10.5.2

$$A = \begin{bmatrix} -2 & 0 \\ 0 & 0 \end{bmatrix}, \quad B = \begin{bmatrix} 1 \\ 1 \end{bmatrix}, \quad C = \begin{bmatrix} -0.5 & 1.5 \end{bmatrix}, \quad D = 0$$

10.6.3

$$A = \begin{bmatrix} -3 & 1 \\ 0 & -3 \end{bmatrix}, \quad B = \begin{bmatrix} 0 \\ 1 \end{bmatrix}, \quad C = \begin{bmatrix} -3 & 2 \end{bmatrix}, \quad D = 0$$

10.8 Uncontrollable.
10.9 Observable.

Chapter 11

11.2

$$K = \begin{bmatrix} 15 & 2 \end{bmatrix}$$

11.4

$$K = \begin{bmatrix} 0 & 1 \end{bmatrix}$$

11.6

$$K = \begin{bmatrix} 1.5 & -2.5 \end{bmatrix}$$

11.8

$$K = \begin{bmatrix} 0.6667 & 0.3333 \end{bmatrix}$$

11.10

$$K_c = \begin{bmatrix} 69 \\ 14 \end{bmatrix}$$

11.12

$$K_c = \begin{bmatrix} 15 \\ -4 \end{bmatrix}$$

11.14

$$K_c = \begin{bmatrix} 115 \\ 18 \end{bmatrix}$$

11.16

$$H(z) = \frac{1.955z - 1.682}{z - 0.8182}$$

11.18

$$H(z) = \frac{(0.04762z + 0.04762)}{(z - 0.9048)} \frac{(0.002268z^2 + 0.004535z + 0.002268)}{(z^2 - 1.81z + 0.8186)}$$

Bibliography

1. Ogata, K. (2010) *Modern Control Engineering*, Prentice Hall, New Jersey.
2. Kuo, B. C. (2010) *Automatic Control Systems*, John Wley, New Jersey.
3. D'Azzo J. and Houpis, C. H. (2010) *Linear Control Systems Analysis and Design with Matlab*, CRC, New York.
4. Dorf R. C. and Bishop, R. H. (2010) *Modern Control Systems*, Pearson, New York.
5. Sundararajan, D. (2008) *Signals and Systems—A Practical Approach*, John Wiley, Singapore.
6. Sundararajan, D. (2021) *Digital Signal Processing—An Introduction*, Springer, Switzerland.
7. Sundararajan, D. (2020) *Introductory Circuit Theory*, Springer, Switzerland.
8. The Mathworks, (2021) *Matlab Control Systems Tool Box User's Guide*, The Mathworks, Inc. U.S.A.

Index

Printed in the United States
by Baker & Taylor Publisher Services